How to Make Our Own News

How To Make Our Own News

A Primer for Journalists and Environmentalists

John Maxwell

Canoe Press
University of the West Indies
Barbados • Jamaica • Trinidad and Tobago

Canoe Press University of the West Indies
1A Aqueduct Flats Mona
Kingston 7   Jamaica

04 03 02 01 00   5 4 3 2 1

CATALOGUING IN PUBLICATION DATA

Maxwell, John.
    How to make our own news : a primer for
    journalists and environmentalists / John Maxwell.
    p.cm.
    Includes bibliographical references.
    ISBN: 976-8125-64-0
    1. Man, media and the environment –
    Jamaica – Handbooks, manuals, etc.
    2. Journalism, Environmental – Jamaica. I. Title

    PN4784.E65 M38   2000   363.7-dc20

Publication sponsored by the Caribbean Institute of Media and Communication (CARIMAC) and the United Nations Environment Programme (UNEP).

Set in Stone Informal 9.5 on 13.5 x 27
Book design by Cecille Maye-Hemmings
Cover design by Errol Stennett

Contents

Foreword

My first task each morning is to wipe down the surface of an old wooden table on the verandah. Afterwards the cloth which I use is black with grime. If the dirt on that cloth is indicative of the pollution in the atmosphere, it is small wonder that respiratory disease is now a major health problem in Jamaica.

Anyone who has to travel to work by bus or car in the Kingston Metropolitan Area of Jamaica sees and smells the pollution in the streets, in the air, in the harbour. Slowly we are choking ourselves to death, not only metaphorically, but literally as well. There is no doubt that we – along with the rest of the world – are approaching the point of no return.

Even our major foreign exchange earner – tourism – which markets Jamaica as a super resort destination of beautiful sandy beaches, clear blue skies and abundant crystal clear waters, is in danger. The report of a critical US agency on the state of Jamaica's resort areas has sent a clear warning that either we clean up or sink in the squalor of our own making.

Recently I had occasion to fly back home from Barbados on a perfectly cloudless day. It was an instructive flight over the islands of Antigua, St Kitts, Puerto Rico, the Dominican Republic, Haiti and Jamaica. It did not need UN statistics to illustrate the scale of the environmental problem facing the Caribbean, one could actually see it with one's own eyes. It was particularly alarming to see the extent of deforestation in Haiti and how fast Jamaica is on the way to repeating that disaster. Yet Caribbean countries are signatories to international legislation. It is certain that the international community will no longer allow us to get away with what we have done in the past. The global nature of environmental concerns puts every one of us – multinational business as well as the individual small farmer and charcoal burner – under international scrutiny. Agenda 21 demands that we clean up our act.

Whatever indices we choose to examine – the state of our children's health, the quality of the food we eat and the water we drink, the pattern of disease and the cost of medical services, the transportation system, the economy, jobs and the financial well-being of our people – all are inextricably woven into the fabric of our environment. If we continue to abuse it, if we do not seriously set about repairing some of the damage, there is going to be a high cost to pay. If we procrastinate any longer, we shall be in the position of Hamlet in the last act of Shakespeare's play: we shall have known what we should have done all along to ensure our own survival, but, having failed to act, we shall have condemned ourselves to certain destruction.

The author of this book is well known for his astute and fearless articles in the national media. For a long time John Maxwell has championed environmental causes. He is in a good position to know the facts, having served on many governing bodies of organisations concerned with environmental matters. In this study he shows the role that the journalist can and must play. Here, he brings together his writing talents and knowledge of environmental issues to present a clear and concise statement, not only of the problem, but of what needs to be done.

While the environment has commanded the attention of writers and scholars in many countries, Jamaicans and others in the Caribbean are woefully short of books and studies about their own situation. It is to be hoped that John Maxwell's contribution to the ongoing debate will be part of the required reading, not only for students of journalism, but that it will also be widely read and inwardly digested by everyone who cares about their own survival.

Dr Elizabeth M. Thomas-Hope
The James Seivright Moss Solomon (Snr) Professor of Environmental Management
The University of the West Indies, Mona

Kingston, Jamaica

Acknowledgments

How to Make Our Own News was inspired by a number of events and developments over several years. Most important, of course, was the Earth Summit in Rio, which marked a decisive change in the relationship between humanity and its native grounds. Rio stimulated the Caribbean Institute of Media and Communication (CARIMAC) to offer special seminars to familiarise Caribbean journalists and media managers with the idea of sustainable development.

Dr Marjan de Bruin of CARIMAC, who has since become my wife, was not only a leading spirit in developing the CARIMAC seminars but was also, with me, the joint author of a course for young journalists employed by the *Daily Gleaner.* This course was the basis for the chapters on writing and editing. Marjan's contribution to conception of this book was therefore crucial. She spent many hours encouraging me to write the book. Later she edited the manuscript and, with David Williams, also proofread it. It is impossible to express adequately my debt to her. The Caribbean Environmental Programme of the United Nations Environment Programme, UNEP, contributed both to the CARIMAC seminars and to the costs of publication. I thank Jan Voordouw, then of UNEP, for his help and encouragement. I must also thank Professor Aggrey Brown, Director of CARIMAC, for his unfailing encouragement and help, and Pansy Benn and Joanne Blake of the UWI Press, for their part in ensuring the publication. To all those who contributed in whatever way, I express my gratitude, while, of course, taking full responsibility for its opinions and for any errors or omissions of fact.

We Can Make the Difference

The eyes and ears and hopes of the world were concentrated on Rio de Janeiro, Brazil, in 1992, as most of the world's leaders gathered there for the so-called Earth Summit, sponsored by the United Nations. The leaders were in Rio, committed to come to an agreement that the Earth and its life forms needed protection against a host of man-made threats.

The world leaders signed an agreement, the Treaty of Rio de Janeiro, better known as 'Agenda 21' – with the stated aim of "reshaping human activities in order to minimise environmental damage" and to provide a programme of action to take us safely into the 21st century and beyond.

What Is Agenda 21?

Agenda 21 is a collection of strategies by which we human beings may redress some of the damage we have done to Earth and to our own chances of survival. Agenda 21 is a global programme for improving the quality of life by caring for our Biosphere, the name given to the thin envelope of air, water and land which supports all known forms of life on the planet. On our lonely voyage through space, our survival depends entirely on how well we treat the life forms with which we co-exist, and how well we husband the slim resources on which we must depend. In doing these things, we will also be protecting the interests of our children and their children.

Sustainable development has been defined as development which does not impoverish the future in order to enrich the present. In short, we must not steal from our children. We need to use natural resources to improve our lives, but we should use these resources in such a way that our descendants will also be able to use them productively.

Think Globally, Act Locally

Agenda 21 advises us to think globally and act locally. Although each of us is an individual, everything we do has an effect on the world around us. Every single thing each of us does has some effect on our neighbours and on the land, water, air and the plants and animals around us. When we take care of our own local environment we are helping to take care of the global environment. As the old English proverb says: "Take care of the pennies and the pounds will take care of themselves."

Agenda 21 suggests that as we take care of our local responsibilities, we should share with each other the news of our successes and our failures, so that others may learn and be guided by what we have achieved. The introduction to Chapter 40 of Agenda 21 says, among other things: "In sustainable development, everyone is a user and provider of information considered in the broad sense . . . at all levels, from that of senior decision makers at the national and international levels to the grassroots and individual levels." This simply means that we all need to learn from each other.

Later in the same chapter is the following paragraph:

> Countries, with the cooperation of international organizations, should establish supporting mechanisms to provide local communities and resource users with the information and know-how they need to manage their environment and resources sustainably, applying traditional and indigenous knowledge and approaches when appropriate. This is particularly relevant for rural and urban populations and indigenous, women's and youth groups (Chapter 40.11).

Abolishing Poverty

Agenda 21 says all countries should work with international organisations, the UN, the FAO, UNICEF and so on, to provide ordinary people with the kinds of information and assistance they need to manage their communities, their work and their lives.

If we are to survive and develop, we must ensure that all of us have a share in deciding and constructing our future, and we must understand that our *major* priority is to abolish poverty everywhere. We need to pay special attention to the poor in cities and in the countryside, and to women and to young people. These groups tend to be bypassed or marginalised in both developed and developing societies.

We should use new technology along with local and traditional wisdom to solve local problems; we should exchange with our neighbours information on how we solve our problems.

Our chances of achieving sustainable development will depend heavily on whether we continue blaming each other for what is wrong with the world or, more sensibly, join together to fix it.

A Defining Moment

"Humanity stands at a defining moment in history."

The preamble of Agenda 21 begins a broad analysis of the problems facing humanity and the things we can do to fix these problems.

The world is threatened by poverty, hunger, ill health, illiteracy and the continuing deterioration of the natural resources which support life on Earth. The situation is dangerous, but we can make our lives safer and more productive if we make sure that development is more in harmony with the living environment. We need to make sure that in the struggle to survive, we don't destroy the plants, animals or any other parts of the environment because their survival is essential to our own survival.

We need to fulfil basic human needs by ensuring that every person has enough food, suitable clothing and decent housing. We also need to improve the general standard of living for all, while protecting and caring for the environment and its ecosystems which make all this possible.

No nation can secure its future on its own; we are all dependent on each other in all sorts of extremely important ways. We need to cooperate in global partnership for the benefit of everybody. All of us can share a safer, more prosperous future if we all cooperate in the movement towards sustainable development.

Sustainable Development

Agenda 21 represents the climax of a long process of global information gathering, argument and consensus. It reflects a commitment, at the highest international levels, to a balanced and integrated approach to environment and development questions.

The primary responsibility for this global partnership lies with ourselves and our governments. This means that every person and every nation must make plans and develop strategies which take account of all the interests around us. Our personal, neighbourhood, national and regional efforts will be linked and coordinated into an international effort by the United Nations (UN). Every country needs to mobilise a truly national effort, involving the government and all the people and institutions of each society.

Sustainable development means many things to many people. Many of us say we are wholeheartedly for sustainable development; but since there are many different ideas about what this means, it is important to understand exactly what it is we are for.

To some people, sustainable development is the equivalent of the sign sometimes seen in some public toilets. 'Please leave this place in the same condition as you found it.' To others, it means much more:

> Sustainable development does not imply absolute limits to growth and it is not a new name for environmental protection; it is a new concept of economic growth.
>
> Sustainable development does not imply a fixed state. It is a process of change, in which economic and fiscal policies, trade and foreign policies, energy, agricultural and industrial policies all aim to induce development paths that are economically, socially and ecologically sustainable. It requires more equitable distribution and equal opportunities within and among nations. It must be a goal for all nations, developed and developing. Indeed, it is a goal for the whole global community. [1]

Sustainable development cannot be simply about leaving the earth in the same condition as we found it. Just by being alive we have altered, changed the world that was here before we were. Over the years, our feet and our hands make small erosions of the surfaces we have touched. They may be insignificant and invisible to us, but each is one more in a long line which began before us, continues after us, and eventually leaves clearly defined tracks or handles polished by use. If we are to achieve sustainable lifestyles we all need to be aware of what we *should* be doing and of what we need to *stop* doing.

Information for Action

Every human being needs to be convinced that he or she has an important part to play in changing the world by changing destructive lifestyles. Every man, woman and child needs to be informed about how to live better, sustainably.

What we need to do – or stop doing – is different for each of us. A poor charcoal burner needs to understand why he cannot continue to cut down trees indiscriminately. A wealthy industrialist needs to understand that his 'disposable packaging' is a threat not only to others, but to his own survival. Both charcoal burner and industrialist need to understand that we cannot indefinitely take more from the earth than the earth can replenish. Each of us needs different information, although all of us need to understand that we – every man and every woman and every child – can make a real difference to the health of the earth as well as to our own prospects for survival.

Motivating People

If we are to practise sustainable development, people need to be motivated to change destructive behaviour into constructive behaviour. If we are to promote sustainable development, we need to have information which will provoke action and convince people to do something that is new and probably different. Since what is new or different to one person may be what is traditional and familiar to another, it is clear that those of us who are activists need to select our messages and fit them to the needs of those whose behaviours we want to influence.

We could, if we had enough money, blanket everybody with the same messages. If we did that we would be contradicting the real messages we want to convey. The blanket or scatter-shot approach is wasteful and may have the unfortunate effect of turning everybody off. We can wage war on attitudes as we have waged war on insects, but that would be just as futile. Like insects, people's attitudes can mutate rapidly in self-protection.

If people feel that their attitudes are reasonable and right, nothing on earth will make them change. They need to understand why what they are doing may not be in their own best interests. Only then is there the possibility that they may change.

A Few Small Fishes

In Discovery Bay, in Jamaica, fishermen were using very small wire-mesh in their fish-traps. Because the mesh was so small, the traps caught hundreds of little fish which had to be thrown away because they were too small to be eaten. The fishermen complained that the fishing was getting worse, that they were having to work too hard to get smaller and smaller catches. They were very efficient fishermen, so efficient that they were fishing themselves out of business. In addition, because there were now so few fish to be viewed by scuba-diving tourists, the fishermen were also losing an important economic sideline.

Fortunately for the fishermen, the University of the West Indies (UWI) had set up its Marine Biology Laboratory in Discovery Bay. The director and other scientists at the laboratory took an interest in the fishermen's problems, partly because their activities were destroying the laboratory's research base and partly out of simple good neighbourliness.

The scientists explained to the fishermen that their fish pots were too efficient. The small mesh pots were catching not only the fish the fishermen wanted to catch, but the next generation – dozens of immature fish were thrown away after every pot was emptied. The fishermen were killing fish which should replace the fish they were catching. By killing fish before they could breed, they were steadily reducing the size of the fish population, reducing their chances of catching larger, more saleable fish and reducing their overall catch.

The fishermen understood what the scientists told them, but kept on buying the same mesh they had always bought. No one wanted to be first to change and put himself at a disadvantage against his colleagues. Eventually, the scientists figured out a way to induce the fishermen to try fishing with bigger mesh. With financial help from a Canadian agency, the scientists offered fishermen enough free mesh to make two fish pots for every old pot turned in.

More than two-thirds of the fishermen accepted the swap. The old mesh was given to their families to make chicken coops. The result: fishermen have reported that they are catching larger specimens of at least one species of fish. And their families have an additional source of income from the chickens, reducing the pressure to catch every fish in the sea.

More important, the scientists and the fishermen are now working on plans to manage the fishery, because the fishermen now recognise that the fish are a resource to be managed and conserved and that the sea is not inexhaustible in its bounty.

The scientists have not so far had any luck with another idea: to declare certain parts of Discovery Bay to be off limits to fishing. They wanted to establish fish sanctuaries where the fish could mature, breed and grow undisturbed. Almost all the fishermen thought it was a very good idea, but no one wanted a sanctuary where he and his father had traditionally fished.

But Not in My Back Yard

This of course, is the hoary so-called 'Not in my backyard' or 'NIMBY syndrome' which prevents many of us taking action where it really counts, in our own local environment.

We are all perfectly capable of seeing what other people are doing wrong, but we look in the mirror only to assure ourselves that we are still beautiful, never seeing our own warts. Part of our problem is that we are often blinded by our basic needs, by obvious, immediate and legitimate self-interest to realise that what we are doing damages our long-term prospects and that what we do as individuals can seriously affect other prople.

Millions of us think that what we do as individuals does not really matter in the overall scheme of things. We are intimidated by forces which we do not understand and do not think we can influence, let alone control. First, we need information to allow us to see where and how we fit in and how we can cooperate with our neighbours to make things better. Second, we need the tools to make use of our information.

We need to understand that each one of us can make a difference, however small. As the Jamaican proverb has it: 'One [by] one, coco full basket.' If we begin to work together, recognising our interdependence – that we depend on each other – we can change the world.

Note

1. Brundtland, Gro Harlem. 1987. "What is sustainable development?" *Sustainable Development*. Panos.

1

Reaching Your Audience

Since most people want to survive and want to live better, information about sustainable development should be ridiculously easy to deliver. People should be thirsting for information, eager to learn as much as they can, as quickly as they can.

The reality is different. Except in the most isolated communities, humanity is bombarded with all sorts of information almost all the time. We are, as it were, in a crowded marketplace, surrounded by vendors all shouting their wares, every one trying to get our undivided attention. The cacophony is terrific, the messages conflicting, the confusion indescribable.

The job is hard. Not many of us always want to do what is good for us. When we do make up our minds to do what is good for us, we find ourselves bombarded by all sorts of messages urging us to do things that may not be good for us, but which are New! Exciting! Fun! And will make us Sexy! Pretty! Rich!

The competition for attention is intense. If we want to get our messages across to our neighbours, we need to penetrate several layers of conscious and unconscious defences before we begin to gain the attention of the audience we want to reach.

These barriers and others like them are useful although they may not seem so at first glance. They help people retain their version of sanity in a noisy, confusing world. Even prejudices are useful to those who hide behind them. Prejudices are short cuts in judgment about whole classes of things which are favoured or rejected purely because they resemble other things.

We may agree that prejudices are bad things, but we need to deal with them and not dismiss them as unimportant.

Similarly, religion, language and education, parts of what is called culture, are useful to us and others in defining personal tastes. They therefore help determine those things to which we will pay attention and those we will dismiss as unimportant or irrelevant.

All of these factors contain additional complications. Class differences in language may vary in importance depending on the context; the philosophies of various religious divisions and sects create sensibilities with which we must cope. One size definitely does not fit all.

Character of the Audience

There are no homogeneous, constant, stable audiences for anything or anyone, anywhere. Audiences change from minute to minute, as people switch on or off (literally and figuratively) and they change for all sorts of reasons.

When President Nelson Mandela speaks to his countrymen as a whole – to the South African nation – about the need for reconciliation, he is addressing a very complex mix of hopes, fears and apprehensions. He is addressing an agglomeration of people which is just beginning to recognise that it does represent a community of interest. When President Mandela speaks to black South Africans, counselling them not to seek revenge for past wrongs extending over three hundred years, he is addressing a different community of interest. The interests and audience are different again when Mr Mandela addresses members of his own party who are both black and white, and who may have suffered, quite recently, at the hands of people who are identifiable and still in positions of power or influence. And when Mr Mandela talks to South African white people as a whole – many of whom support him – he is addressing an audience which is again different from people who are members of the formerly all-white National Party, most of whom – now including some blacks – have been and are his political opponents. But in every case, he is speaking to a South African audience.

Injecting the Message?

Earlier in this century, when the science of communications research was young, many advertising practitioners and others in communications thought of audiences as passive receptacles into which the communicator poured or injected his ideas. Audiences were thought to be more or less homogeneous communities of interest, all marching to the beat of the same drummer. More recent research has demonstrated that audiences are a vital part of the communications process; that audiences have infinite skills in transforming messages directed at them.

Audiences are not homogeneous communities, but are made up of individuals with their own ways of looking at the world. These world views are formed by many different influences: parentage, upbringing, religion, education, personal experience, prejudice, ignorance, fear and all sorts of other subjective and objective factors. Even twins brought up under apparently identical circumstances may react differently to the same stimuli.

The same story has different significances and resonances for each of us. We each take from the same text what we consider important to us and by doing that, we, the audience, confer importance or particular meaning to any text. In crude language, we can say that audiences hear what they want to hear. Audiences in fact, define the message. We can go even further, to say that audiences use the media for their own purposes.

We want to communicate with people in order to influence their behaviours. To do that, we therefore need to operate within the same general context as most of our audience.

We need to speak to our audiences in language that is transparent to them, language they themselves understand and can speak. The ideas and metaphors we use must be ideas and metaphors with which they are familiar. We need to speak about aims and aspirations which most of us share.

Above all, we need to make ourselves worthy of their trust by telling them the truth as we understand it.

Although some of us have never given the matter any real thought, every person must have an interest in his or her environment. We tend to behave in ways which (our training and experience tell us) will work to our advantage. Sometimes we cannot see that what appears to be to our advantage in the short term may actually be against our real interests. Remember the fishermen of Discovery Bay?

Transmitting information to people, making them more aware, increasing their knowledge, does not mean that they will change their behaviour. They will only change if they see the prospect of an advantage in changing, if they can see a better life beckoning.

Journalists, politicians, businessmen and the church are all trying to reach mass audiences. Each has his own version of the truth and each is convinced that his message is of paramount importance. As we remarked earlier, the marketplace is crowded, the messages cacophonous, conflicting and confusing.

Getting the Message Across

How can we make our messages count in this babel? How can we get people to listen?

Getting the sustainable development message across will be easier if you are trying to reach people who have heard of sustainable development, who have heard of Agenda 21. If your community is conscious that some things are wrong with the Earth environment you, at least, have somewhere to start. Unfortunately, most of us have to start at square one, talking to people who may have heard of 'the environment', but don't really understand how it affects them and how they affect it.

To reach the mass audience, the mass media are normally the prime means of delivering the message. For decades, baby harp seals were beaten to death to make perfect white fur coats. No one cared, until someone brought a television camera into the proceedings and people everywhere could see and be revolted by the obvious cruelty of the hunt. Similarly, until César Chavez and his United Farm Workers began the boycott of California grapes and lettuce in the 1980s, few people were concerned about the use of pesticides on the food they ate or worried about the working conditions of the people who reaped their food.

Both of these issues were important in producing environmental consciousness in millions of people. Neither would ever have been heard of outside a tiny circle had they not been reported on by the mass media. In one case (grapes) the media found the event, in the case of the seal hunt, the event found the media.

What Are 'the Media'?

The communications media generally are said to include:
Traditional, mainstream media:
- Newspapers, magazines
- Radio and television
- Film

Alternative formats in mainstream media:
- Call-in shows on radio and television

Alternative media
- Theatre, including folk theatre
- Billboards, posters, newsletters
- The postal service, the telephone system, etc.
- The Internet

Each medium requires a different approach. Each approach depends on what the medium is capable of doing and the audience you wish to reach. There used to be a wonderful billboard a few miles from where I live. Its message:
"Want to learn to read and write? Phone: 123 4567"
Not many people realised at first glance why this beautifully produced advertisement was utterly useless.

Reach of the Media

In my country (Jamaica), when times were gentler and slower, there always seemed to be people on shop 'piazzas', as we called them, reading the newspapers aloud for people who could not read. These days, most people who can't read are too ashamed to admit it, so the day of the public newspaper reader is over. Since newspapers must be read to be understood, the only people who will get your message through the newspaper are those who can read.

In most countries most people get most of their news and other current information through the radio. The transistor has made everybody a consumer of information transmitted by radio. This means that many people who cannot read or write know quite a lot about world affairs and about what is happening in their own countries.

Television is a powerful medium. Unfortunately, television sets generally are too big and heavy to carry around so most television is watched at home.

Radio and newspapers are both indoor/outdoor media and can take their messages anywhere.

Newspaper information is in a permanent form. It can be put down today and read when and where the reader wants to deal with it.

Newspapers and television deal in both words and pictures, they show and tell, and each can provide excellent opportunities for teaching.

Newspapers are static media. Their pictures don't move, their words need physical exertion to be read. But newspapers can also print diagrams and pictures and can be excellent to teach readers, for instance, how to make a solar cooker or a rammed earth wall.

Television viewers don't have to do anything but look and listen. Radio audiences can work while they listen and are least tethered or limited in their movement by their medium.

Radio and television can transmit news of an event as it happens. They are great for allowing vicarious public participation in important events. Millions of people round the world watched the first moon landing on television and millions more have watched Michael Jackson concerts, the Olympics and the World Cup.

The medium you choose depends on what sort of information you wish to transmit, who you want to get the information, where they are and when you want them to get it.

Use of the Media

Newspapers are great for getting wide coverage among people who read. Still, many people who can read do not read newspapers. To reach them you may need to try other media.

Magazines normally reach an audience that is generally somewhat better educated, more sophisticated, more affluent than the people who read only newspapers.

Newsletters are usually produced for special audiences about special subjects. If the audience and the information are both carefully selected, newsletters can be very influential and cost-effective media.

Television is watched by most people above a certain income level and by many people below it.

In many countries, a great deal of television programming comes from abroad and some stations are hungry for local material. Some television stations would love to be able to run locally made documentaries about local questions. Some have news features or magazine programmes in which they deal with local activities. You may be able to suggest people they might interview, subjects they might care to investigate. Local television newsrooms are always looking for interesting footage and people who do interesting things. If your local station has nothing on sustainable development or Agenda 21, call up the programme director and find out why. It may simply be that no one has ever suggested it and he has never thought about it.

Radio reaches everybody. Some people listen only to music, some only to news, others spend their time following soap operas or call-in shows. Somewhere in your part of the world is a radio station which will be interested in what you are doing. If there isn't, why not try to see whether you can raise the funds to start a community radio station?

Radio call-in programmes are probably the cheapest and most effective way to get people to hear about important matters. If there is a call-in programme in your area, one of the best ways to find out the local level of interest in anything is to call up and discuss your interest.

At its crudest, you may call the programme to ask why do the people of X-ville have to put up with a garbage dump next door to them? You may get more of a response than you imagine. You can always tell the host the name of your organisation and explain how interested people can get in touch with you.

Recruiting support

If you are lucky, people will call in to agree with you, perhaps to give a history of the problem and to suggest what could or should be done to fix it.

If you haven't got an organisation yet, a call-in show may be one way to recruit people to discuss forming one.

Perhaps your local radio station may be interested in starting a call-in programme on sustainable development – even once a week – and you may be the person to do it. Don't worry about whether you have the right kind of voice for radio – your audience will be much more interested in your information than in the timbre of your voice.

Exhibitions and displays of all types can be used in schools, libraries and other public places to educate people about sustainable development and to display some of the problems of your community. Invite public figures to go on guided tours with you, so that you can point out things that need to be done.

Make sure that the people of the community know about what you are doing so that they can add their knowledge of the situation to yours. Talk to your local authorities and representatives. If things don't change, complain. Better still, get together with your neighbours to fix them if you can. If you can't fix what's wrong, organise to put pressure on those who can.

Newsletters, which you can publish yourself, provide an excellent means of publicising your activity within your own group and outside. Make sure that everybody in your organisation gets one, and send them to local editors and to members of your local authority and to anyone who you think should know about what you are doing. Newsletter publishing is not as difficult now as it used to be. Today, almost anybody can produce newsletters that are presentable.

When you can afford to, get a computer like a Macintosh iMac for about US$1,000. The ClarisWorks programme will probably come free with the machine and you are in business as a publisher with the addition of an inkjet printer like an Epson Stylus or a Hewlett Packard DeskWriter. The whole setup shouldn't cost more than US$1,200 to $1,500.

Campaigns for the achievement of some particular goal can be useful to energise people, raise morale and allow others to see that what you are doing is good and may be emulated. Campaigns may be small or large, aimed at entire populations or discrete sectors. The form of the campaign will depend on your objectives and resources.

What Do you Want to Do?

What are your objectives? What do you want to achieve? Is what you want what your community wants?

The mass media may not be the answer for what you want to do. You need to use the mass media only if you want to reach a mass audience. You want your community to know what you are doing and how they can play a part. An exhibition of photographs, a guided tour of environmentally hazardous or questionable sites, or even T-shirts printed with your message may be more appropriate for your particular purpose.

Some organisations are so inefficient that they specialise in sending out news releases which never get published. If you want to be noticed and taken seriously make sure that your work is useful, that it benefits the community and is what the community wants.

Publicity

When you have planned your activity, it is useful to issue a news release to the community press or to call them in for a cup of coffee and a chat or to go to see them to explain what you are doing

If you need larger audiences than the neighbourhood or regional media can deliver, you need to plan entirely differently. But, before you start planning, you need to justify to yourself and your group your need for a mass audience. You may find it makes more sense to begin with a more limited set of objectives and depend on public interest to make it bigger.

In all of your plans, you should make sure that everything you want to accomplish has a completion date. Make sure there is someone responsible for managing that aspect of your programme. Set targets and make sure you meet them.

All mass media will pick up and carry news *they* consider important. What you consider important may not be what the news media consider important. If their ideas of what is important is different from yours, perhaps you might try to change their minds.

There are two main ways to go about changing their minds:
• Make yourself and your cause so important to so many people that you cannot be ignored;
• Become known to the media.

Despite everything you may read in this book and elsewhere, what is important in any newsroom is what the editor in charge of that newsroom considers important.

These days there are not too many editors like the famous (or infamous) former owner of the *Chicago Tribune*, Col. Robert McCormick. For nearly thirty years the crusty and prejudiced old Colonel used his newspaper to reflect his personal likes and dislikes, his prejudices and his enthusiasms, and kept out of the paper news about developments with which he did not agree. People like Col. McCormick are not too common these days; nevertheless, it is wise to try to ensure that publishers and other media managers understand who you are and what it is you are trying to do.

The conventional media may be useful for some projects, not so useful for others, but it is always useful to have them on your side. They can, on their own, expand and amplify your reach and your influence if they think that what you are doing makes sense for their audiences.

If your organisation is a community group, start off by finding out whether the news media have any correspondents in your area. If they have, invite them to your meetings. Explain what your group's objectives are, who is eligible to join (everybody, one hopes) and how you intend to achieve your aims.

Whenever you are starting any project, it is always a good idea to begin with a short list of achievable objectives. Grand plans may look and sound good, but unless you are an organising genius, many of your objectives may fall by the wayside because there is just too much pressure on too few people to get even the routine things done.

Since this book is not about organising a community group, there is only one thing more to say on that subject. That is, your group should discuss publicity for your cause at every meeting you have. You should take note of what is being said about your group, gently correct mistaken impressions, if necessary, and invite any local representatives of the media to your meetings. You have nothing to hide and everything to gain by making your activities accessible to the public. You are, after all, discussing public business.

With your group organised and various people in charge of specific things, it is a good idea to decide on a spokesperson for the group, and, if the group is big enough and is involved in several projects, to have spokespersons for each project, with one person who can speak on behalf of the whole organisation.

Sustainable development is everybody's project, and it is likely that if you take time to explain the ideas behind "Think Globally, Act Locally" you will attract many willing volunteers and later entire communities, who want to play their part in making the world healthier, starting with your neighbourhood or community.

If you are one of those people who likes to tell everybody what is good for them you may have a problem. Leadership is much more about articulating the good things that people want and helping them organise to achieve their goals.

Below is a checklist for organising your publicity programme.

Organising Publicity: a Checklist

Who (what community/interest group) do you want to reach?

What is the size of your target audience?

Where do they live?

Where do they work?

What sort of work do they do?

Why are they likely to be interested in what you have to say?

Is your message likely to be thought important by them? If not, why not?

How will it affect them?

How will you know whether your message has reached most of the people you want to reach?

What kind of response do you hope to get?

What kind of response do you expect to get?

How will you measure the response?

What do you do next?

When you have answered those questions you should have a good idea of how to go about organising publicity for your cause or your group.

Keep your planning simple and make sure that anyone who has any part to play knows what is expected of him/her.

Make sure people understand the importance of being on time wherever they have to be.

2

Dealing with the Press

This chapter is not about writing, it is about being written about – about how environmentalists, scientists as well as others, can establish themselves as sources of information for the media. It is mainly for non-journalists, particularly for people who need to deal with journalists and need to establish their credentials with them. I hope to give you some insight into the kind of approach journalists take towards their sources.

When we make up our minds, when we decide to do something, we may be acting on instinct – on a hunch or a 'gut feeling'. More often, we look at the choices we have and make decisions on the basis of what we know. We try to evaluate the information and give more weight to what we think is reliable and less weight to what we think is less reliable. Common sense, you may say.

Using common sense, one of the ways we tend to judge reliability is by the reliability of the source of the information. We may have two sources for our information: one source has been more often right than wrong, while the other has been more often wrong than right. Some people, we know, are great rumour mongers, ready to pass on any story if it is juicy enough, regardless of whether they know it is true. Others are more guarded and probably less likely to pass on false or misleading stories. The credibility of a story depends very much on the credibility of the storyteller.

Journalists judge their sources in the same way and are trained to pay attention to people who are not likely to mislead them.

Some of the people reading this book may not want to write their own stories, but may have information which the public should know. You may, for instance, be the representative of a non-governmental organisation involved in sustainable development. You may want to get information out to the public by way of the mass media. If so, you need either to make your own news or to get someone to make it news for you.

Becoming a Source

Specialists can be specialists only if they know their subjects. But while one specialist, Jacques Cousteau, for example, could be endlessly fascinating talking about what he knows best, some experts are not only boring, but may even give the impression that they don't really know what they are talking about.

How can an expert not be on top of his subject? Simple. He knows so much about it that he doesn't bother to wonder *what it is that other people might want to know*. He talks around the subject, never getting to the point, or talks above our heads, or disguises the point in specialist language, so people become confused and/or lose interest.

Normally, when most of us want to tell a story, we introduce it by saying something like 'Do you know what happened?' That provokes the listener's curiosity and also promises to satisfy it in a short time. And most of us can give a quick summary of something that has happened, of the important points of any event, in less than a hundred words.

We need to cultivate and refine that common sense ability to tell a story. Our natural talent will be our most important asset in getting and keeping the attention of the people we want to listen to us.

In telling a story to a friend, we usually try to be as factual and accurate as possible. We want to be believed and we usually want our friend to know as much as we know. We take our friend into our confidence, as people say. And when our friend knows something we don't, we believe we will get the 'real' story, knowing that what we hear is the truth or as close to the truth as we are likely to get.

So most of us usually try to tell our stories as factually as we can and as completely as we can. This is a good principle when we want our stories to be published outside of our personal circles.

If you want to get your story into the news, you must know it better than anyone else. You must also be able to convince journalists that you

are reliable – 'a witness of truth' as they say in the law courts.
Journalists tend to trust people who:

- Know what they are talking about
- Are straightforward, having no hidden agenda
- Are generous in sharing their knowledge
- Are willing to explain exactly why they cannot say more than they need to
- Are neither overbearing nor ingratiating.

Who Are you?

Are you a scientist, an activist, a concerned citizen or just an interested bystander? A journalist needs to know what sort of interest you have in a story so that he can let his informed judgment and his gut feeling tell him whether he should listen to you and trust you.

At the outset of any relationship with journalists you need to make clear exactly who you are, why they might find it useful to listen to you and why they should believe you.

If you are an interested party you need to make that clear – quickly. Explain what your interest is, so that the journalist can understand that he or she is only getting one side of the story. You can always deceive a journalist or anyone else, once.

Don't play games with journalists; it isn't worth it. Your long-term losses will almost certainly outweigh your short-term gains.

If you are an expert, say so. You don't have to boast about it, just make it clear that you do know what you are talking about. If you are indeed an expert, you will find that journalists, if they trust you and find what you say useful and/or interesting, will come to you of their own free will to ask your opinion or solicit your expert knowledge.

Speaking with Authority

There is an important benefit in disclosing your partisan interest: if journalists want to get 'both sides of a story' and if your position is known, they may call you up to find out what you think whenever there is any movement in an issue that they think may concern your interest.

If you want to be accepted as an authority on a particular subject, you should be able to demonstrate that you are an authority. If you happen to be a professor of zoology, for instance, most journalists will accept you without question as an authority in the field of zoology. Sometimes reporters may ask you to speak on scientific or other questions outside of your speciality. Some reporters will do this because they trust your judgment, others, simply because you may happen to be the nearest available scientist.

If you have an informed opinion on any subject, don't be afraid to give it, but be sure to make it plain whether you are speaking as an expert in the field or simply as a scientist who is interested or maybe, even more simply, as a concerned citizen. The fact that you are a scientist does not make you an authority on all matters of science.

A concerned citizen who is also a professor tends to carry more weight than a concerned citizen who isn't a professor, and some people are easily tempted to throw this weight around. Be careful; your credibility capital is built up slowly; it can be destroyed overnight. That said, you should have no fear of sharing your opinions if you think you have something useful to say. You simply need to be careful that you say what you really mean, unambiguously. And you need to make sure your reporter understands exactly how much weight to give to your remarks. If not, your words may be taken out of context because of a reporter's ignorance or inattention or, more rarely, by malicious intent.

Journalists are taught to coddle and preserve their sources. If you are a reliable source, the serious journalist will respect your integrity as you respect his and may come to depend on you.

Your qualifications are important so far as they tend to prove that you are what you say you are. Who you are – your character and reputation – is another matter. If you speak as a concerned citizen, all you have to prove is your concern. If you are an environmental activist, it may be helpful to have available a brief resumé of your recent history. Don't be offended if you are asked to prove who you are. Reporters need to be able to assure their editors that their sources do exist and are likely to be reliable.

Honesty Still the Best Policy

Most human beings find it difficult to deal with people who don't seem to be quite what they say they are. A wolf in sheep's clothing is feared, a sheep in wolf's clothing is ridiculed. People are much more comfortable with people who are straightforward. If you have an interest in any public issue and you want to have any public influence on the matter, declare your interest up front.

People tend to take a dim view of 'independent' voices who later turn out to have been on the payroll of some interested party. When the time for exposure comes, as it almost always does, the process tends to be brutal, shaming and damaging to the pretenders and to their sponsors.

It is not easy to continue to live in a community which does not trust you. It is impossible to have any real influence if you are not thought to be trustworthy. It is tragic to watch a reputation, built up over many years of hard work, vanish overnight because of one ill-conceived stratagem. *You cannot fool all of the people all of the time and some people you can't fool at any time.*

Summary

If you wish to have influence in public affairs:
- Be yourself. Do not put on an act
- Say what you mean. Be sure of your facts
- Take the media into your confidence
- Do not conceal or disguise your interest
- Honesty is the only intelligent policy.

3

Story Ideas from Agenda 21

When we make our own news, we are trying to do several things: We are informing and educating people about their environments and their vital interests in the health of their environments. In addition:

- We are informing them about their right and their capacity to manage their own environments in their own interest and in the larger global interest;
- We are guiding our communities towards finding and developing the resources to develop themselves sustainably.

Sustainable development begins at home. Since we need to be sure that our own houses are in order before we begin to set the world to rights, we can use this exercise for another purpose: we can use it to ferret out the kind of news stories which will help inform people about the state of their environments and the state of their world. In other words, finding out whether our own houses are in order is a good way to develop awareness about the quality of our environment and the news which we can make out of Agenda 21. A good way to start is to conduct an environmental impact assessment of our own houses and our own lifestyles.

What sort of impact do we individually have on the rest of the world? On other people? Have you ever considered how much impact you personally have on your country's national resources? It is a good idea to make up a checklist to make sure that we don't overlook anything. Here are a few areas you should look at:

- If you waste water, for instance, you are among those who are ultimately responsible for water being rationed or locked off during droughts. Think about it. If we multiply each of our households by all the houses around us, we have a lot of wasted water to talk about, to write about. We have a story.
- Do you waste electricity by leaving lights burning?
- Do you drive when you might walk?

There are dozens of areas of your own life which you should check for wasteful, unsustainable use.

Many of us feel quite helpless to affect the course of our lives. We feel that our problems are so small that no one 'in authority' will think our problems worthy of notice. But, if we discover that our problems are not unique, that our problems are part of much larger situations, we will find it easier to bring them to public and official attention, and make it more likely that they will be noticed and corrected.

Looking carefully at our own lifestyles may lead us to discuss important questions such as consumption patterns: Do you consume a high proportion of imported goods? Does your country? Why is this so? Are there no substitutes available locally? What is the effect of this consumption on your country's balance of payments? Does this kind of consumption have any other bad effects? This is the sort of story which has enormous implications for sustainable development. It is also the sort of story which needs to be very thoroughly researched and carefully handled. What you may think is non-essential may seem vital to someone else.

The large issue of unsustainable consumption patterns is made up of many small components. Almost every big issue starts small, and you can track most of them from your own house, your own street.

On the Street Where You Live

When you are satisfied with the state of your own house, look at the condition of the street outside.

The Street Itself
- Is the street cleaned regularly?
- Can people walk safely on the sidewalks or do the sidewalks need repair? Are there any dangerous obstructions to pedestrians?

- Is the road surface in good repair? Are there any obstructions to the free flow of traffic? Is water flowing in the street? What is its source? Is it sewage or some other waste? If it is waste, find out what is in it. Why is it flowing in the street, instead of being properly treated and disposed of?
- Are there any dangerous or misleading signs on the street? If there are overhead power lines, are the trees regularly pruned so that there is no contact between trees and electricity? Whose responsibility is it to make sure the trees and the lines are safe? Are these things true of other neighbourhoods? Of other regions? Of the whole country?
- Where do the children play? Do they fly kites which may get entangled in power lines? If they play on the street is there any alternative space for playgrounds nearby? Does any work need to be done to make the playground safe for children? Can you and your neighbours do it?
- Are there any abandoned cars, other vehicles or discarded appliances such as old refrigerators on the street? These can be very dangerous to children. They are called 'attractive nuisances' and should be removed to be disposed of safely, and not simply dumped somewhere else. Old motor vehicles contain dozens of toxic substances, from plastics and PCBs to corrosive acids and other materials.

Water   Make sure that you do not waste water, that there are no leaks in your system. What's the source of your water? Is the water safe? Is it treated before it gets to you? How is it treated? Does the water look clean? Is it colourless? Does it have any strange taste or smell? Do children in your area suffer from stomach ailments caused by water-borne diseases? Gastroenteritis, dysentery, cholera, and typhoid are some examples. Your local health department should be able to inform you about your neighbourhood and about the wider community around you.

- Can you get your water tested for contaminants like bacteria, sulphates and nitrates, petroleum pollution, pesticide pollution, fertiliser residues, lead?
- If your water comes from wells, are the wells located near to a waste dump or sanitary landfill? Is the dump or landfill really sanitary, or do wastes leak from the landfill into the groundwater?

Child Health   Are the children in your neighbourhood sickly? Do many children have stomach trouble or ailments such as asthma? What is the infant death rate in your country? If it is more than 30 per thousand your country has a serious problem. What causes it to be so high? Ask doctors and people in the health services about what causes these problems and what can be done to correct them.

Sewage Do you know what happens to the sewage from your house? Does it flow into a sewage system? Is it treated in any way? Where is it discharged? Most sewage is not properly treated to kill disease organisms and much of it is simply pumped into the sea, polluting the sea water, stimulating algae growth, killing corals and poisoning marine life in general.

- Is this what is happening to your sewage? Is your community aware of this?
- In many communities in the developing world, sewage simply goes into pits in the yard. In many communities, even in the developed world, sewage is simply pumped into the rivers and the sea.
- Minerals, chemicals and bacteria in sewage can contaminate the groundwater and end up in drinking water taken from wells or rivers nearby. In the state of Florida, where cities pump their sewage into deep wells, there is now real fear that sewage contamination is threatening the freshwater aquifers.
- If your sewage is pumped into the sea, what are the effects on the fishermen and other workers near the dumping sites? Do people swim nearby? Are they aware of the danger? What are the effects on marine life, corals and the like?
- You may find that dumping sewage at sea has several other serious effects on human health and on the health of the environment. Try to find out if anyone can put a money cost on these effects – on the damage to coral reefs and beaches in lost fishing and employment and in terms of diseases; schooldays and workdays lost to illness.

Solid Waste If you have yard space, you may be able to turn your food wastes, if any, into fertiliser by composting. Go to the local library and read about composting. It is a simple process which can provide manure for your plants and healthy exercise for you. Could some of your community's garbage be composted? What would the benefits be?

- Try to separate your garbage according to type; paper, glass and plastic are recyclable. If these materials are not recycled in your community, find out why not.
- For solid waste the message is: Reduce, Reuse, Recycle. Reduce the

waste you create by avoiding wastefully packaged goods, and find ways to reuse or recycle whatever waste you produce.

Air Quality Do many children in your neighbourhood suffer from asthma or other bronchial ailments? What is the incidence of asthmatic and other lung diseases among children and adults in your community, in your country? The health department should be able to provide the statistics.

- What are the atmospheric pollutants in your area? What processes cause them? Who is responsible? Are there laws or regulations to control atmospheric emissions? Are any being broken? Agenda 21 says that polluters must pay to clean up their pollution.
- Do you live on or near a farm or near a factory? Is the air clean? Is there a smell of burning or any other strange smell in the air? Find out what causes it and how it can be controlled. Your family's health may depend on it.

Pesticides In most parts of the world, powerful insecticides, many based on highly toxic chemicals derived from nerve gases, are available from supermarkets and local shops. Are there any laws to govern the importation and use of insecticides in your country? Are these laws enforced?

- If you live on or near a farm do you know if chemicals are used to kill weeds or to kill insects?
- What chemicals are used? Are these chemicals dangerous? Are they being used legally and according to the instructions? Who supervises their use?
- Are there any signs of chemicals in your water, in the atmosphere?
- Do you sometimes see dead birds in your neighbourhood?
- Do you notice any odd smells, can you 'taste' the air?
- Have many of the ordinary insects – moths and butterflies, for instance – disappeared from your area? Have you noticed whether there are fewer birds than there used to be? If so, watch out – you and your children may be at risk.
- Find out what is happening, inform your neighbours and discuss what you need to do to stop the dangers and the nuisance.
- Find out from the nearest university, or from chemistry teachers in schools, whether the chemicals used are safe and sound. What damage can they cause? If they are not safe, you have a story.
- Is there any official organisation which regulates commercial stan-

dards in your country? What are the standards for insecticides? Do the insecticides on sale meet those standards? Make the same inquiries for herbicides used on lawns and in agriculture.

Forests As people become poorer, there is increasing pressure on forests for fuel and for food. What is the state of your country's forests? Are they being cut down for firewood or to make charcoal?

- In some places, forests owned by the nation – by you – are being harvested illegally. If there is a sawmill near you, do you have any idea whether the timber processed there has been obtained illegally? If you suspect that the trees are being stolen, report your suspicions to the police or to the department responsible for forests.
- How much forest is being destroyed every year? Deforestation causes soil erosion, producing landslides, sending tons of soil into rivers to clog up dams and disrupt water supplies, sends masses of soil into the ocean, killing coral reefs and starving the fish.
- What are the effects in your country? What is the cost? What is the government doing? What can the local communities do? Has anybody investigated the planting of so-called 'charcoal forests' or 'fuel forests' – specially planted and managed stands of trees which can be regularly harvested for fuel?

These are a few examples plucked randomly from chapter headings in Agenda 21. In each chapter there are many, sometimes dozens, of other story ideas. If you do not know what something means, ask. You may find you have a good story.

Story ideas may also be found in searching for examples of people who have found their own solutions to sustainable development problems.

A Village Built on Garbage

In Cairo, Egypt, the people of the village of Mokattam make their living from garbage. The village is next to a huge dumping ground for all kinds of waste. Many people began to pick through the garbage looking for things to sell or reuse. Volunteers from outside the village decided to help them.

The men and children collect the garbage by night and the women sort the garbage by day. From the waste, they make compost (fertiliser and soil conditioner) and pig feed. The villagers are recycling paper by hand and sewing quilts while they learn to read, write and develop other marketable skills.

With the help of volunteers and foreign assistance, Mokattam has established several schools, a health centre and a centre where the women learn to weave and sew. The volunteers enlisted several volunteer groups, including university students, to go into the more affluent areas of Cairo to carry the message to the householders who are producers of the garbage. If the primary producers can be convinced to do most of the primary separation of their garbage, it will mean that less waste will be contaminated and therefore unrecoverable.

The volunteer group which started the revolution in Mokattam is called the Association for the Protection of the Environment. Despite its grand name and its great achievements, it consists of nine middle-class women. Their work has helped raise living standards among the 20,000 people of the village, reduced the infant death rate and increased the self-respect of everybody involved.

There are similar examples in many other parts of the world. In Trinidad in the West Indies, Beetham village near Port of Spain is, on a smaller scale, an example of the same kind of effort as Mokattam.

4

Developing Story Ideas

Agenda 21 makes it plain that sustainable development requires a very wide range of action by people everywhere. Whole societies need to be involved, and this means that information and public education are vital to sustainable development. Everybody needs to know what he or she can do to make sustainable development a reality.

Agenda 21 and the arrangements for implementing it will provide us with headlines as well as a blueprint for community action. Some of us may combine both, making news out of the work we do. Making news this way helps other people realise that they too, can make the difference in their communities and even outside. We need to understand that real development, sustainable development, does not come from abroad, from outside, but from satisfying the real needs of real people, from supplying the needs of people just like us.

Unfortunately, there is a serious problem with Agenda 21: although it is a document which affects the life of every person on this Earth, nowhere in the world does the document itself seem to be freely available. Agenda 21, 40 chapters plus the declarations, contains over 180,000 words, about 600 pages.

One of your priorities may be to make sure that the sense of Agenda 21 is available to your community. If it is not, you may want to make it one of your projects to make sure that the ideas in it become available to everybody. One way to do this is to make it your business to publicise

Agenda 21. Perhaps you may want to set up a group – a non-governmental organisation or NGO – to publicise Agenda 21 in your community, in your country.

Unfortunately, no officially authorised condensation exists which would make Agenda 21 more accessible to ordinary people. That is why this book includes a summary of the document (Appendix 1), giving the outlines of the actions proposed but not going into the detailed arrangements. UN offices can provide the document on computer diskettes and these offices are the first places to go if you want to see the full document. Another version which may be useful is a precis for journalists, published by Panos, an international NGO in Washington. Another short summary, _Agenda For Change_, is published by the Centre for Our Common Future, in Geneva, Switzerland.

An Agenda for Everybody

If you want to help involve your community in working for sustainable development, a good place to find out how to go about it is chapter 36 of Agenda 21. This chapter explains how to promote action on sustainable development from community level to international level.

Most of the world's governments agreed to the terms of Agenda 21, promising to implement it in their countries, but it is a sad fact that in many places there has not been much action. Among the reasons for this lack of action is the fact that, in many places, there is no single person or body responsible for ensuring compliance with the undertakings in Agenda 21.

Although many people want to be involved, most people don't know how they can get involved. We need to tell people about Agenda 21 and explain to them how they can play their part in making the world a better place for themselves and their children. Agenda 21 is above all about personal responsibility. It is about developing an ethic of survival which encompasses all living things.

Much of the news you read and hear consists largely of progress reports of one kind or another. Even a report of a murder trial, for instance, is a kind of progress report, although it may contain, from day to day, arresting new facts which tend to rivet the imagination and may seem to be new stories. In the same way, progress reports on Agenda 21 do not need to be dull.

Every new advance, every additional school or institution which becomes involved in Agenda 21 is news to somebody. We do not need to seek for the transcendental events – these will happen anyway. What we need to do is concentrate on the enormously appealing and useful human stories about how ordinary people, doing ordinary things, accomplish extraordinary transformations in their own lives and in the lives of others around them.

News from Communities

One of the major aims of Agenda 21 is to spread a general public awareness that everything we do affects the environment in which we live. Public awareness is part of a global education effort to allow people to understand and analyse what is wrong with their environment and to work out ways to fix it.

The more people know about Agenda 21, the more likely it is that they will want to become involved in promoting their own programme for sustainable development. So, first, we have to make people aware of what Agenda 21 means to them, we need to make people understand what they themselves, their communities and their governments can do and should be doing.

Increasing Public Awareness

What is the state of public awareness of Agenda 21 in your country, and what action is your government taking to implement it?

If you are a member of an environmental NGO, you probably already know. If your government is actively pursuing the objectives of Agenda 21, most people in your community and your country should be well aware of what is happening.

If, however, there seems to be no action on Agenda 21, you need to find whoever is responsible for directing your country's Agenda 21 programmes and get the story. Find out why nothing is happening. Since we must presume that all governments signed Agenda 21 willingly and enthusiastically, if things are not going well it is likely that the government itself is not properly aware of what is happening or of what needs to be done. In many countries political leaders are so caught up in the day-to-day problems of the society that they find no time to look at the larger picture. If your government cannot explain how it is dealing with Agenda 21, you have a very good story and you should be able to get it published.

How do we go about finding out?

First, we need to know which organisation or person in the country is responsible for overseeing Agenda 21. If, for whatever reason, you don't know, the obvious place to find out is from your Government Information Office. If they don't know, try the ministry or department which deals with agriculture. That office is almost certain to know and may very well be the place where you will find the people responsible for the programme. As a last resort, and if all else fails, other places to try are the local office of any UN agency, or any foreign embassy. Since foreign embassies deal in technical cooperation they should be able to identify who they deal with in your government on Agenda 21-related matters.

When you find out who is responsible, the first question to ask is whether your government has ever published any progress report on the status of Agenda 21 programmes in your country or your community. You are entitled to this information.

When you get the status report on Agenda 21, check to see if it is up to date. If the last activities reported are more than, say, six months old, find out why there is not more recent information. It may simply be a question of the mechanism by which reports reach the responsible office.

That too may be a story.

You need to talk to the person responsible for compiling the status report. He or she should be able to tell you about the progress of Agenda 21 activities in your country – whether or not things are moving at a good pace.

Grassroots Decision Making

Agenda 21 depends heavily on transferring decision making authority to the local level, to the people and communities directly affected by development. It is important to discover how far your country or region has moved in this direction.

By 1996, local authorities everywhere should have undertaken consultations with their communities to draw up local plans for implementing Agenda 21. The communities should have been assisted to produce their local versions of Agenda 21 – their own local plans for meeting the challenge of sustainable development into the next century.

The story: Did the local authorities do what they were supposed to do? If they did do their jobs, where are the local plans? Are the local authorities aware of what they were expected to do?

- If they were not aware of these responsibilities can the government explain why they were not aware?
- What are the plans, if any, to start the consultation process?
- If there are no plans, why are there no plans?
- If there are plans, when will the process begin? Will all communities be consulted? If not all, what proportion will be consulted and why not all?

Procedure: The first part of this story is easy. Call up the manager or secretary of your local authority (town council, parish council, etc.) and find out whether she or he knows about this need for consultation. If your authority is not aware, find out how many more are similarly ignorant.

- Ask your local authorities if they have heard of Agenda 21. If they have, find out what they know.
- Then go to the next level of government, to the ministry or department which supervises local authorities. Do they know about this responsibility?
- If they do know, ask them what they have done and what are they planning to do. Dates, places, projects and procedures are important in any plan. Find out as much as you can.
- If the people supervising local authorities do not know about their responsibilities, go to the next level, which is probably the minister or secretary of state.
- Ask him whether he is aware of what is to be done.
- What is he doing or planning to do?
- By the time you have got this far, you should have a good series of stories about the real state of action on Agenda 21 in your country.

Talk to the People

A good way to expand this continuing story is to go out into selected communities and find out what the communities want. The best way to start is probably to attend existing community groups, such as parent-teachers' associations, farmers' organisations and youth clubs to get some sort of consensus of what the people think their communities need and whether they are aware of Agenda 21.

You can ask them about educational opportunities for children and adults, the status of women in the community, their main economic activities, access to health care, land use and other questions which should suggest themselves when you visit the community. And, of course, you can ask them if they have ever heard of Agenda 21.

By this time, if your stories have been published, they should have gained a fairly wide audience, and sustainable development should be a live issue in your community. If you have been scrupulously accurate and fair, you should not find it difficult to widen your coverage and to advance the cause of sustainable development. Every chapter of Agenda 21 can produce such stories. You can get leads from the summary included in this book and see what Agenda 21 actually says in detail about any subject.

For instance, Agenda 21 recommends that the knowledge and science of indigenous people should be given more serious attention, because there is every reason to believe that within these cultures are benefits unknown to the rest of us, including medicines, for instance, and 'new' foods. In any case, and for their own sakes, the people and their cultures, should be helped to maintain their intrinsic worth, their dignity and their integrity.

Cultural Integrity

In many parts of the world, people who live in special situations are becoming conscious of their value, not as exhibits, but as equal contributors to the whole stream of human experience and culture. Agenda 21 envisages special measures for their protection and development. Because their culture has been valuable to them, many of them are now beginning to realise that there is not much future in being mass producers of baubles and trinkets for the tourists of the world.

They are trying to protect their cultures from being transformed and degraded – from being their own to being 'exotic' or merely 'quaint' – leaving them with nothing to sell and nothing to show for what has been sold. They feel in danger of losing something more valuable than all the money they could make by indiscriminate exposure.

An Ethic of Survival

In the Caribbean island of Dominica, the population is very conscious of its relatively unspoiled environment and the people are working out ways to allow visitors to enjoy their landscape, their wildlife and their environment, while keeping it unspoiled. Indigenous people – once called 'Carib Indians' – who met Columbus when he arrived 500 years ago, still live in Dominica. This group of Dominicans is devising plans to finance their own development by allowing visitors, on a carefully monitored basis, to come into the ancestral lands to meet them, buy their handicraft and learn about their culture. The indigenous people are going to great lengths to make sure that their second major encounter with the rest of the world is not as destructive as the first, when they were overrun by the Conquistadors.

They do not intend to exhibit themselves as inhabitants of some sort of human zoo. They expect to meet their visitors on equal terms, although they each live very different lives.

Not 'Going Native'

Since they live in a distinct part of the island, the indigenous Dominicans are making plans to ensure that they themselves, their culture and their environment – the landscape, the mountains and the forests, the trees and animals, rivers and pools, many of which have great cultural and religious significance – are properly respected and not prostituted and ruined by being turned into a kind of living theme park with themselves as exhibits.

They are, of course, under enormous pressure from cruise lines, travel agencies and even some other Dominicans, to open themselves up to exactly that sort of treatment in the interest of making lots of 'money for everybody'. Having lived without lots of money for centuries, they seem to be in no particular hurry to embrace some of the more demeaning aspects of modern life and tourism.

Tourism can be a powerful force for education because it exposes people to new and sometimes valuable experiences which are otherwise inaccessible. The indigenous Dominicans are aware that too often, tourism has become a means for reducing the host populations simply to a kind of parasitic existence in which they expose themselves and their culture for the delectation of people who are looking for thrills and don't care how they get them. The indigenous people want to know whether materials they have used for centuries for medicines, for instance, and for other purposes may be valuable to the rest of the human family.

Many indigenous people possess traditional knowledge and technology once thought to be irrelevant but which are increasingly valued in a world more and more limited by its dependence on a few resources. A report on underexploited tropical plants makes the point:

> Throughout history man has used some 3,000 plant species for food; at least 150 of them have been commercially cultivated to some extent. But over the centuries the tendency has been to concentrate on fewer and fewer. Today, most of the people in the world are fed by about 20 crops – cereals such as wheat, rice, maize, millet and sorghum; root crops such as potato, sweet potato and cassava; legumes such as peas, beans, peanuts (groundnuts) and soybeans; and sugar cane, sugar beet, coconuts and bananas. These plants are the main bulwark between mankind and starvation. It is a very small bastion.[1]

The report goes on to explain that although many indigenous species possess as much merit as many cultivated crops, they were disregarded during the colonial era when consumer demand in European countries largely determined the cultivation (and research) priorities in tropical agriculture.

> Even after independence, the pattern of concentrating on a few crops changed little. Markets abroad were established, and the new countries needed foreign exchange. Furthermore, as indigenous scientists were generally trained in the institutions of temperate zone countries, they had little interest in studying tropical species. Even the food preference of local populations in tropical colonial countries became so influenced by European food habits that in many places local demand for traditional crops declined.

One casualty of this process was a plant with enormous potential for increasing protein production in protein-poor tropical countries. In many countries the leaves of the amaranth are used as a spinach substitute. In Jamaica, for instance, it is called kalalu or callalloo. What most of us do not know is that the grains in the seed-heads of the amaranth family are rich in protein and exceptionally high in lysine, one of the critical amino acids usually deficient in plant proteins. The lysine level in the edible amaranth is higher than in high lysine maize varieties and is about the same as in soy meal.

In the fifteenth century, the Spanish church suppressed the growing of amaranths in its efforts to eradicate 'pagan' Aztec religious ceremonies in which amaranth was an essential element.

Amaranth grain is usually parched and milled and the dough used for 'pancakes', cooked for porridge or popped like popcorn as a confection. It can also be powdered and made into a drink.

No wonder the Aztecs worshipped this amazing plant. Yet, 500 years later, most people in Central America and the Caribbean are ignorant of its worth, except as a pot herb. We import wheat, corn and soy meals and flour when we have a superior product which grows in any back yard.

Note

1. *Under-exploited Tropical Plants with Promising Economic Value.* Report of an ad hoc Panel of the Advisory Committee on Technology Innovation, Washington, D.C., National Academy of Sciences, 1975.

5

What Is News?

More of us are increasingly becoming aware that *everything* we do influences our chances for survival. That concept may be clearer to the subsistence farmer and to the beggar in a downtown street than to the mogul or office worker in a glass and concrete office tower, but it is no less real for any of them.

Every time someone uses an old-fashioned aerosol spray, he or she is contributing to the destruction of the ozone layer. Every cigarette is additional pollution in the atmosphere, every soft drink bottle thrown onto the roadside degrades the quality of life for every one of us.

Agenda 21 recognises that if we are to achieve sustainable development, everything that each of us does is important and each of us is important in our own right. Each of us is both a consumer and a producer of information. Small things may be just as important as large things, ordinary people just as important as the so-called newsmakers. If the world is to change, it is we who must change it.

When we write or broadcast on the subject of sustainable development, we are talking, essentially, about making the world a better place for everyone, about helping to increase everyone's chances of survival. When we talk about aerosols and chlorofluorocarbons (CFCs), we should realise that we are also talking about our own prospects of getting skin cancer, for instance.

Why Does It Have To Take So Long?

Armed with the faith of the true believer, some of us will try to bludgeon people over the head with survival issues. We *know* that we have an audience which is fundamentally interested in survival, so we go to it and are surprised when people pay no attention. We *believe* we have good strategies for survival, things that everybody should know and practice, yet we find it difficult to convince our next door neighbours – our natural audience – that what we are saying makes sense for them.

The truth is mighty and shall prevail, we are advised, but why does it have to take so long?

At this point it may be important to remember that we can help straighten out the world, as long as we don't try to hammer every individual into shape. People need to understand that we are not talking about vague, airy-fairy concepts, but that we are talking about their future happiness, their children's welfare and all sorts of things that add up in the end, to plain self-interest – survival by doing what, strangely enough, is good for you.

Unfortunately, many environmentalists actually provoke hostility to the cause they represent, because they behave as if they alone know the sacred mysteries of survival and sustainable development. They are intolerant, not only of opposition, but of ignorance and even of innocence.

We need to remember that like us, most people resent know-alls and, perhaps more important, many people are embarrassed to admit their ignorance about something as important as their environment.

Sometimes, in our enthusiastic arrogance, we upset politicians and others by behaving as if small advances or small improvements are worthless. We need to welcome every move in the right direction, no matter how tiny.

Human beings are by nature conservative, slow to change. Inertia keeps rocks in their places, and a great deal of energy may be needed to move one. Once it has been moved, much less energy is required to keep it moving.

Taking It Easy

We need to be gentle, persuasive and tolerant.

Many people don't understand much about 'The Environment', but they do understand that their lives are affected by some natural and artificial forces.

If people are to understand how they can make their lives better, they need to understand why life is not as good as it could be in the first place. That is where we have to start.

Some people have been born into an existence which hardly allows them to dream, to imagine that things could be different. Others are born into environments where their parents and families take privilege for granted and don't seem to realise that what they consider to be simply satisfying their impulses may seriously injure somebody else. In the United States, for instance, many people have embraced the 'frontier life', living in isolated settlements in the mountains. For many, that lifestyle is not as idyllic as it may once have seemed. According to a *New York Times* story a few years ago:

> For some mountain inhabitants, wood stove heaters are a matter of survival. Many depend on wood as the sole source of warmth in regions where natural gas is unavailable, conversion to propane is costly and electric heating would be more expensive than the rent.
>
> But for others, wood smoke is a source of discomfort and disease that can force them to stay indoors for days. 'I don't go out of my house unless I have to,' said Janet Glover, a 52-year old Quincy native who blames her emphysema and asthma in part on a lifetime of breathing wood smoke. 'If I open the door, it hits me. I feel like I'm imprisoned in my own home.'

This story is from the wide open spaces of the American west. Nearer home, some of us burn garbage in our back yards while others live near to dumps which catch fire from time to time, releasing poisonous gases and noxious particles into the air we breathe. Every person has individual needs; one size does not fit all. As the old cliché says: One man's meat is another man's poison. One man's barbecue is another man's asthma.

We need to understand that sustainable development is not a single recipe which can be applied to every situation. Sustainable development consists of what works best in particular situations.

When we write or broadcast about sustainable development, we need to remember that the most learned among us was once ignorant, that life is about learning and that we can teach most effectively by understanding the needs of those we want to teach. This applies to journalism just as much as it applies to teaching in a classroom or preaching in a church.

So before you begin to write your story, think carefully about what you have to say, why you think you need to say it and, above all, try to imagine what it will mean to the people who will see it, read it or hear it.

What Is News?

Why do we find some stories more immediately interesting than others? Why do some stories grab and rivet our attention? Why do we remember some stories for months and even years, while others seem to vanish instantly into a grey unfathomable limbo?

If you think about these questions you may come to the conclusion that some of the decisive factors involve you, yourself, and the things that interest you, and also that some decisive factors involve the person who tells the story and the way the story is told.

It is impossible to explain why we remember some things and not others, why a silly song persists irritatingly in memory, while some important facts melt like ice-cream on a hot pavement.

Journalists talk about 'news values' when they try to account for the varying levels of interest by particular people in particular stories.

News Must be New and Interesting

For news to be News, it must be new to the person who is receiving the information.

People are not normally excited by old facts, stuff they have heard before. Last year's catalogues and yesterday's newspapers are garbage to most people. They need to know 'What's NEW!' and NEW is perhaps the most potent single word in the advertiser's armoury. But what you may consider old hat may be brand new and very important to someone else.

Apart from its newness or newsworthiness, information is interesting for many other reasons. It may be strange, outlandish, it may be inspiring, offering an example to emulate or avoid; it may be funny, making us laugh, or tragic, moving us to tears.

Stories may be about heroic deeds or squalid machinations, about new scientific discoveries, about natural or man-made disasters, about dinosaurs or outer space. All interesting stories have one thing in common: they appeal to our humanity, to our emotions, especially to the feeling that by knowing a little more, we are better equipped to face the world on its terms.

When people can identify their own interest in a story, the more interesting and important that story will be to them. You may ask: How can people identify with a story about dinosaurs? How do they identify with stories about pulsars and galactic disasters millions of light years away? How do they identify with stories about Marilyn Monroe or the Yanomami 'Indians' of the Amazon? With stories about Mother Teresa in Calcutta or gorillas in the Central African forests? Think about it.

News Must be Relevant

When you take a story idea to an editor, the first question you are likely to be asked is some variation on "Why is it important? Why should I be interested?"

If you are asked this question, what sort of answer would you be able to give?

The first factor to consider is: How many people (readers, listeners) is the story likely to affect and how is it likely to affect them?

Will it make a significant number of people happy, pleased or satisfied, or will it make them sad, apprehensive, fearful or frustrated?

If the number directly affected is small, is the news sufficiently dramatic or significant in some other way to interest many people other than those directly affected?

People will consider news relevant, interesting or important, because the news –

- affects them personally or people nearby, their neighbours and friends;
- concerns people they know or know about;
- is about dramatic human events, conflicts, quarrels, murders, wars, love affairs, dramatic rescues, terrible accidents;
- is about dramatic natural events, hurricanes, earthquakes and droughts;
- is about the organisation of their lives, government action, for instance, new regulations, new laws;
- is of emotional interest, provoking sympathy, disgust or pity;
- may be so odd, so strange or inexplicable that it makes them wonder, marvel or speculate.

What Is a News Story?

We have been looking at news, so far, from the layman's point of view – very generally. Let's look at it from the professional journalist's point of view.

It is not very different from what you might expect, not very different from the commonsense judgment of news. But the journalist has a somewhat sharper focus. Let's look more closely at how news decisions are made in newsrooms.

Every story is, in one way or another, about people – either about individuals or groups of people, or about mankind and the fate of humanity. Stories about the stars in their galactic courses are just as much about survival as is a story that Mrs Reagan consulted astrologers. A news story is a living entity which never stops growing, however slowly, until it dies. Some stories are stillborn, others grow to great size and age.

News is in the very air we breathe – sometimes it is poisonous enough to kill, as in Bhopal in 1987, or in the seawater in which we swim; in the rain, as in the Mississippi floods of 1993, and the wind, as in hurricane 'Gilbert' of 1988, or in the lack of rain, as in the droughts in the Sahel desert. Real news is on the streets, everywhere. It may even be in your backyard.

What makes an event worth reporting – worthy of a reporter's attention?

What Makes News?

There are two generally accepted definitions:
- News is information about some change in, or interruption of the 'normal course' of events, something unexpected. News must be new.
- News is information people need to make rational decisions in their lives. News should be useful in some way.

But how does a reporter decide which events are unusual enough to be 'news' and what information it is that people need to know?

Certain guidelines have emerged from practice and are used to define 'news values' or newsworthiness generally. These factors may operate separately or in combination. They include:

Significance   An event may have effects or repercussions on the lives of a significant number or class of people. Journalists talk about the 'significance', 'importance', or 'weigh' of events and this significance is related to the range of effect that an event may have on the lives of people.

Everyone is affected by a rise in taxes; some people may be more seriously affected than others. People need to know the degree of *impact* that events or developments are likely to have on their lives.

Will they be better or worse off ?

What's the reason for the change?

Sometimes the effects of an event or of a development are not immediately obvious. Certain changes in government policy or the weather may not seem very important when they happen, but may have far-reaching effects or consequences on people.

Unusual/Phenomenal Events   Are of course news. Events or behaviours that are strange, out of the ordinary, uncommon, unexpected, unique or bizarre are, really not rare. Earthquakes, hurricanes and other natural phenomena may precipitate disasters, but then again they may not. It is only when they have an impact on human life that they become important.

Some years ago, off the coast of Iceland, a volcano suddenly erupted in the sea, forming a new island. The development of this volcano and this island held the world's attention for months. The idea of a volcanic island being born in front of our eyes was sensational.

Mass murderers are phenomenal news, as are people who have too many wives or husbands. When a woman gives birth to triplets, it is not big news; if a cow did, it would be. If something is strange, unusual or is not known to have happened before it is likely to be newsworthy.

Timeliness   Sometimes called 'immediacy'. André Gide, the French novelist, defined journalism as "Everything that will be less interesting tomorrow than today." Freshness, newness, unexpectedness is the stuff of most of what is defined as news.

Prominence   Well-known people, places, institutions or ideas influence newsworthiness and make people want to know more. When your next door neighbour smashes his car it is probably not news, except to his family and the people who saw it happen. If the Prime Minister or President has the flu it will be news, but, sadly, it probably won't be news if your neighbour even dies of pneumonia.

Sometimes reporters use the excuse of a person's prominence to report trivia as news, and sometimes, "people are famous for being famous", as Andy Warhol is alleged to have said. Journalists must use their discretion to decide whether an event is really news or whether they are reporting stuff that has no real significance simply because it concerns somebody famous or notorious. Trivia is the curse of the reading classes.

Proximity   People are interested in things that happen near to them. Three people being murdered in Jamaica will be bigger news in Jamaica than six people being murdered in Barbados, other things being equal. In Barbados, the opposite will be true. But if three Jamaicans are murdered in Barbados, it will probably be bigger news in Jamaica than if the same three Jamaicans had been murdered in Jamaica.

Conflict   Events that reflect extreme differences or clashes between people, institutions or ideas may make news, because they appear to represent a diversion from the 'ordinary' workings of civilised life, or because they symbolise the human struggle for survival.

People may quarrel or fight with each other, may injure or kill each other, may struggle against adversity, may attempt to dominate their environment or may in some other way come into conflict with their fellow man or with the forces of nature and so make news.

Human Interest   Stories make news. People are moved to laughter, to tears, to outrage, to sympathy, to wonder and to pity at stories about how ordinary people cope or don't cope with extraordinary events.

Human beings are social animals and most of us can easily imagine ourselves being in someone else's shoes and are affected by things which happen to other people, even to animals. People are 'touched' by things that happen to people they have never met, even by people and situations which they know are fictional, as in TV serials.

Topicality   Current events and developments that are being talked about may be 'commonplace' in real terms, but are still important to many people, and may make news. The Arab-Israeli Peace process is not a single event and we don't concentrate on it all the time, but many of us are interested in the process and the events which are part of it – the advances and the reverses, even in the events which don't seem to make sense.

As things develop they may seem to gain or lose importance from time to time; there may be new revelations against a background of familiar material. Stories may begin with great impact, and over time lose the initial drama, but may still contain news of interest to important segments of the population.

Stories may start small, and suddenly grow more important, as more and more facts are unearthed or because some other story may amplify the effect of the first. The worldwide epidemic of AIDS, conflict in the Balkans, the increasing degradation of the environment are all examples.

Although any one of these factors is sometimes sufficient to make news, more often, two or more of these factors in combination push stories into the news. Some people think that impact and strangeness are the keys to newsworthiness. Others suggest it is timeliness and prominence. There is no magic formula, and there are other factors which influence the worth of a story, which suggest to a reporter or an editor whether the story should be covered, how it should be covered and in what detail.

Audience   Also determines newsworthiness. What is news for a local newspaper is not necessarily of any interest whatever to a national newspaper. An evening newspaper may find a story, say, about local musicians much more newsworthy than may a morning paper published in the same city, but appealing to a different audience. Space considerations sometimes influence 'newsworthiness'. Stories which won't get into a 24-page paper may find their way easily into a 48-page paper.

Policy   Of the news organ influences the judgment of newsworthiness. The newspaper which describes itself as a 'newspaper of record' will publish material which will not have the slightest chance of getting into an afternoon tabloid. And afternoon tabloids publish stories which would never be considered for publication by certain other papers.

Pressure   From publishers, advertisers, governments, political parties or other forces may colour or tend to colour the judgment of reporters and editors. News is sometimes suppressed or magnified because of the prejudices of the publishers or because of pressure from other sources.

The news 'mix', as it is called in North America, influences the *volume* of news carried about certain classes of events. Sports and entertainment news are gaining ground in many newspapers at the expense of 'hard news'. Some editors feel they must carry specific quotas of foreign news, of regional news and community news, and some papers are departmentalised to reflect these decisions.

Competition   Determines to some extent what news is news. In economics, Gresham's Law of Coinage says: "Bad money drives out good." A kind of Gresham's Law appears to operate in journalism, and a newspaper which raises its circulation by specialising in scandal will often be imitated by its competitors because they believe that they need to 'fight fire with fire' to retain their readers or to gain new readers and a larger advertising base. Things which were not considered newsworthy become 'newsworthy' because someone decides that they may help sell newspapers.

Analysing Newsworthiness: a Checklist

In analysing the probable appeal of any news, particularly news about the environment, about the human condition, we have several criteria by which to judge:

Is the news 'good' or 'bad'?

What makes the news 'good' or 'bad'?

How significant are the good or bad effects?

Is anyone in danger?/Will anyone benefit?

Who is in danger?/Who will benefit?

How serious is the danger?/How much good will be done?

Does it affect your audience personally, their families and friends, their neighbours, people they know, people they know about and are concerned about?

Will some people or class of people be more affected than others?

Does it affect their chances for survival?

Will it mean lifestyle changes for anyone?

What sorts of changes?

For whom?

Does it affect their jobs, their houses, their health, their wealth, their comfort or general feeling of well-being and security, their short-term or long-term prospects?

How soon will the facts in your story begin to affect identifiable people or classes of people?

Are some people already affected?

Do they know that they are already affected?

When will the effects begin to seem significant to the people affected?

If the effects are not good, can anything be done to avoid or postpone them?

Is the problem caused by natural forces?

By human action?

If by human action, who is responsible?

What, if any, damage has already been done?

To whom or to what?

Can any estimate be made of the extent of the actual and potential damage, either in terms of people, area or money ?

Can the situation be corrected? If so

How can it be corrected?

By whom can it be corrected?

Is there any time limit for the correction process to begin or end?

How much will it cost in terms of effort, manpower, time and money?

If the situation cannot be corrected, why not?

What other options exist?

What are the cost and/or benefits of these options?

Will accepting these options improve the general welfare?

Can those affected live with the problem?

If they can live with the problem, should they?

If they should not, why not?

If they must put up with the problem, what will it mean to their over-all chances for survival, comfort, prosperity, general welfare?

Is there any reason to hope? Is there any cause for optimism?

I am sure you can think of many other kinds of questions. Naturally, few stories can answer every possible question and in many cases there is no need to cross every *t* and dot every *i*. *The major issues in every story are the things you need to be concerned about.* By going through the checklist above you will quickly be able to sort out the major points of your story. Some of the minor issues may really be quite unimportant.

But, go through the list above again and again, just in case you may have overlooked some significant factor which may lead to a new development. Sometimes, factors which seem insignificant contain the potential to become extremely significant.

You should realise that any story which leaves important questions unanswered is neither complete nor will it be satisfying to the reader. Your reader or listener wants and needs and deserves the information. Rationing news is unfair, dangerous and fatal to credibility.

6

Writing News Leads

Before a reporter ever begins to write a story, he or she must understand, as clearly as possible, what the story is about. If you are not a journalist, you need to make sure that you have a close focus on exactly what you wish to communicate to your audience.

You must know what you have to report. You need to organise the information to make sure it will be read by as many people as possible, most of whom, you hope, will want to act on the information in your story.

Newspaper readers need to know the basis of a story quickly. They want to know why a story should interest them. If that question is not answered quickly, they will turn away to something else.

If you want your stories to be read, you will try to get the reader's attention immediately and to hold it until the story is finished. It is possible, but not usually necessary, to capsule a story in one sentence or in one paragraph. What the reader needs right away is to gather the essence of the story, the central issue.

Leads In General

The lead is the most important part of a news story. It acts as an invitation to the reader. It follows the advertisement provided by the headline, and should serve as an overture to a spirited recital of interesting facts.

A good lead gives the essence of the story, promising to deliver in more detail the facts the readers feel they must know to be well-informed.

Leads should use simple language to make the essential points as quickly and cleanly as possible. Sentences should be dynamic, supple, muscular – not flabby or fussy. The reader must get a feel for the action from the words, from the effective use of language.

The sentences should be declarative, stating clearly what they are about. The voice should be active, relating the movement of the event – who *does* what; not passive – explaining what *has been done* to whom.

Actions should, if possible, be described from the point of view of a dispassionate onlooker, interested, but not personally involved.

Provoke Interest

Ideally, leads should be straightforward, concise and clear. A lead should whet the reader's appetite for more. It should therefore not be an exhaustive and boring catalogue of everything that happened, just enough to get the main thread of the story onto the reader's agenda.

When the lead is good, the reader will feel that he ought to find out more. The lead should, therefore, take advantage of any element of drama present in the facts, because human beings like to be entertained as well as informed.

This does not mean that the reporter should try to force drama into a story. He must deal with the facts. If the facts are dramatic, as they often are, it makes sense to preserve the drama but not to overplay or downplay dramatic elements.

Let's Write a Lead

Let us suppose that in the city of Kumina, certain people keep cutting down the trees on the Arawak Hill which is a prominent feature of the capital city. Below the hill, on Mountainside Avenue, there are many houses, some of them very expensive mansions. There are also many smaller houses in a squatter settlement nearby.

The Geology department at the university has studied the problem over several years. They tell you that Arawak Hill is becoming geologically unstable because of soil erosion. The soil has been eroding because people have been cutting trees for charcoal and squatters have burned land for small cultivations.

There is now a danger that if the clearing of the land continues at the present rate, within about ten years almost the entire mountainside could collapse after one of the heavy rainstorms to which the area is subject. Such a landslide would bring millions of tons of mud and rocks down onto the houses at the bottom of the hill. Many lives might be lost, especially if the landslide happened at night.

Damage could run into millions of dollars. Mountainside Avenue roadway would probably be buried under the debris, and there would be major economic disruption since the road is one of the most important traffic routes leading out of the capital city. The city is also the chief port and main manufacturing centre.

The scientists say the hillside can be made safer if people stop cutting trees and start replanting. If tree cutting continues at the present rate, it will become increasingly dangerous to live in any of the houses on Mountainside Avenue.

With those facts, let us try an experiment.

Try to select the most important elements in the story as I related it to you. Number them in the order of their importance. Now, in a hundred words or less (about ten lines of writing), set down the major facts as if you were relating the story in a letter to a close friend who is temporarily abroad, but who owns a house in the area.

When you have done that we will see how well you have done. First, let's examine my story. What is the single most important fact? Is it

- The possibility of major economic damage caused by the destruction of hundreds of houses?
- The fact that many lives may soon be in serious danger?
- The possibility of a major landslide?
- The cutting down of trees for charcoal and land clearing?
- The possibility of major disruption of the transport system leading to major economic loss?

Although all these factors are important, the one that would concern most people – and people are your audience – is the danger to human life and whether that danger can be averted. Your friend would be concerned about the safety of his family first, his neighbours and his property next.

That, in my view, is the important fact on which we must concentrate.

Facts *au naturel*

Beyond the facts as we know them are other factors, which every editor must take into account because he is not only responsible to his news organisation but also to his readers.

How should the story be treated?

While many people are in danger, they are not in immediate danger. They should be informed about the proximity and extent of the danger so that they can take reasonable steps to avoid it. Even if they were in immediate danger we do not want to cause a panic.

We have established the fact that the hillside can be saved; disaster is not inevitable. That is important. But we need to make sure that, first, everyone understands that there are very real dangers and that, second, everybody is given the time and the information which will allow them to take whatever action they think is appropriate.

Some newspapers will seize on the most sensational aspects of a story like this one and will deliberately set out to cause the kind of excitement which can lead to panic. We might see something like this:

ARAWAK HILL TO CRASH

THOUSANDS WILL DIE

DAMAGE IN THE MILLIONS

Arawak Hill is on the move and scientists predicted yesterday that when the hill collapses it will bury thousands of people, destroy hundreds of houses and cause millions of dollars in damage.

Geologists at the University said yesterday most of the houses on Mountainside Avenue, both the mansions and the shacks, could be destroyed by a massive landslide from neighbouring Arawak Hill.

According to Professor Franklin MacDonald of the University's Geology Department, most of Mountainside Avenue could be destroyed by a major landslide within the next five years.

The geologist said that such a landslide would not only kill thousands of people, but would block the main highway out of the city's east end, causing tremendous economic disruption. Prof. MacDonald said the danger is caused by the illegal felling of trees on Arawak Hill, making the hillside unstable. So many trees have been cut, he said, that the hillside could slide at any moment, bringing millions of tons of mud and rock down. The scientists said the Big One may not happen for another five years, but small landslides are likely at any time, threatening both the mansions and the squatter houses below the hill.

We want to warn people, not frighten them out of their wits, so our approach would be different. We would try to make the report as un-sensational as possible, stating the facts clearly, not trying to cause unnecessary alarm and distress, but making plain that people urgently need to take action to protect themselves and their interests.

As you will see, the difference between a sensational story and a sensible story is a matter of nuance and emphasis. You will see how just a little touch (a 'gloss') here and there, can alter the whole character of a story.

Our lead would contain the same elements of the story as the sensationalist version:
Who is threatened – By What – Why, How and When – Who says so – What can be done.

SCIENTISTS WARN OF
ARAWAK HILL SLIDE
Tree-planting can save
lives and property

People living on Mountainside Avenue have been warned that they have about five years to protect their houses from destruction by massive landslides.

Professor Franklin MacDonald said yesterday that Arawak Hill is becoming dangerously unstable because charcoal burners and squatters continue to cut down the trees on the hillside. The University's top geologist warned that unless action is taken quickly, almost the entire hillside could collapse within five years, burying dozens of houses and killing hundreds of people.

The scientist said indiscriminate land clearing was destabilising Arawak Hill. Small landslides are already beginning to cause minor damage to squatter houses below the hill.

Prof. MacDonald said there could be a major disaster if the hillside is not sta-bilised by intensive tree planting. If tree cutting continues, a major landslide was likely within about five years. Such a landslide would destroy hundreds of houses, probably killing hundreds of people if it happened at night. It would also be an economic disaster, because traffic out of the city's eastern, manufacturing end, would have to find alternative routes to the port.

"It doesn't have to happen", the geologist said. Prof. MacDonald recommended an immediate and massive tree planting programme to stabilise the soil on the hillside.

When we examine this story in terms of its news values we find that it has several of the attributes which make news.

News Values	Facts
Impact, significance	Many people in danger
Prominence	Extensive property damage Economic disruption
Phenomenal Event	Major landslide
Timeliness	Work must start now
Conflict	Economic interests of squatters, charcoal burners against householders' and other economic and human interests
Human Interest	The whole story
Audience	The story is local

Of all the news values we talked about, the only ones missing from the story are competition and pressure. Not all stories are so packed with newsworthy facts.

A reporter confronted by a story like this may find it difficult to decide just where to start. Others find it easier: they head straight for the 'people factor'. But there is more than one 'people factor' here. First there are the lives in danger; second there is the economic survival of the squatters and the charcoal burners. If you deal with the lives first you have a ready-made follow up: what options exist for the displaced squatters and charcoal burners?

How Stories Work – the Five Ws

The body of a story is defined by:
- a skeleton of fact – what happened (or is likely to happen);
- fleshed out by the people – who did what or to whom it happened;
- enclosed by a web (or skin) of circumstantial detail or context – when and where;
- clothed by the causes or motives which may have brought the event about – why and how. The reader needs to know the Who, What, When, Where, Why & How (the so-called 5 Ws) of the story or to get a pretty good idea of those facts, quickly.

If the reporter's mind is focused on the essence of the story, he will be able to summarise what happened clearly and succinctly in one or two sentences. The most important or interesting factor may be, as we saw in the Mountainside story, any one or a combination of 'Ws'. Whatever it is, you should write your lead around the central factors.

Writing Effective Leads

There is no formula for a lead. Sometimes *Who* is involved may be most important. If a well-known person (or place or thing) is involved, that fact may be important enough, unless another element is more important. A *Who* lead may be justified even if the person is not himself well-known but is related to someone or some fact that is well-known.

Mr Peregrine Falcon, brother of Prime Minister Gyr Falcon, died in a motorcycle accident yesterday.

What may include things or events. Events are sometimes more important than the individuals involved.

Molasses, twenty sticky tons of it, gummed up East Street and stopped traffic in downtown Kingston yesterday. Police said the heat wave had caused molasses to ferment and overflow from a huge storage tank at Water Lane and East Street.

Where may sometimes be important enough to overshadow the other Ws, especially if the place adds extra resonance to an otherwise routine story.

Last Street, Denham Town, was the end of the road for Hezekiah Campbell, one of Kingston's street people. He was found dead there yesterday. Police say there was no foul play.

One factor which is often overlooked is the *Why* element – the motive, force, reason or dynamic factor behind an event.

> Malachi Campbell wanted to see his mother. She lives in Canada and Malachi had no money to travel. So, two days ago he tried to stowaway on a ship loading gypsum for Canada. Yesterday, Malachi Campbell's battered body was found in Kingston Harbour. Three crewmen from the ore-carrier *Atlantis* have been charged with his murder.

There are occasionally instances where time – *When* – is important but rarely is it the most important feature. It usually becomes important because of its connection with another interesting fact.

> Less than twenty-four hours after he was acquitted of the charge of murdering his father, Lester Mallester was arrested on a new murder charge.

The *How* lead is not often used, because it is not always easy to explain the mechanics of an event or an action simply and concisely enough for a lead. It works best in a so-called delayed lead, where the facts come out less than obviously.

> Last week it was a piece of cake, but that didn't work. Yesterday it was a meatloaf, and that worked.
>
> According to the police, a hacksaw baked into a meatloaf was the instrument used by four men in last night's escape from Central Police lockup in Kingston. Last week police confiscated a file inside a piece of cake brought for one of the prisoners. Yesterday, inside the cell, another file and the remains of an half-eaten meatloaf told the rest of story.

Reporters need to be sure they understand the real significance of an event or a fact.

The *What* must be clearly significant to the audience or its significance must quickly be made obvious. Similarly, a reporter who leads with a name must be sure that the name means something to a significant number of his readers, and the reporter who chooses time or place or cause for the lead must be able to convince his editors and his readers that that was the best way to start.

The lead in the next story is not at all obvious. It works because the lead is so unusual and un-newslike that, hopefully, it intrigues the reader into wanting to know what exactly is happening:

> Jimmy Edwards won't be moving house again.

Edwards moved to Kingston ten years ago when Hurricane Allen destroyed his house in Port Antonio. He moved house again, this time to Florida when Hurricane Gilbert destroyed his house in Kingston eight years later.

Yesterday, two years after Jimmy Edwards' latest move, Hurricane Andrew destroyed his house in Miami, Florida. But Jimmy Edwards won't be moving house again. The day before Hurricane Andrew destroyed his house in Florida, it killed Jimmy Edwards when it sank the boat he had chartered for a fishing holiday off the Bahamas.

As I said earlier, there is no formula for a lead. Effective leads depend on the writer's vision, imagination, on the connections she or he can see in the basic facts.

Writing For Radio

Writing to be heard is very different from most writing that is meant to be read. A *reader* can always go back over a sentence or a paragraph that is not clear. The *listener* cannot do that.

When someone listens to a story, as fact succeeds fact, his brain must be able to assimilate everything as it happens. If the listener does not understand one reference or has to puzzle out the meaning of a difficult phrase or even one word, he may easily miss the next thought or even the next sentence or two and the entire story may become meaningless.

The key to good writing for radio is that it should be simple, clear and easily understood. It should never try to pack more than one new fact into each sentence.

Sentences should be short and uncomplicated.

Writers must understand that time does not allow them to tell the story in all its detail. A clear, simple outline of the most important facts is all you can reasonably expect to deliver.

A radio news story to be read as one item shouldn't be longer than about 100 to 120 words – about ten lines of typing. A story of 100 to 120 words takes about 60 seconds to read and that is about as long as most people are willing to concentrate on any item of news from a disembodied voice on a loudspeaker or a talking head on a TV screen.

Most stories should not really be longer than seven or eight lines, about 80 words or 45 seconds. Of course, important stories with many important facts, in a narrative form, are acceptable if they are important enough. Too many radio news editors bore their listeners with stories which are too long and too complex.

In some places, as in much of the United States, editors go to the opposite extreme, producing stories which are so short that they are finished before the listener really understands what the news is about. Recorded quotes, the so-called news-bytes, now last about 10 to 15 seconds. Such butchery of the news is inexcusable and defrauds listeners of their right to be properly informed.

Of course, radio and television news may be illustrated with 'actualities', that is, audio or video recordings inserted into the news.

7

Writing the Rest of the Story

Having composed a good lead, the next problem is how to tell the rest of the story. If the lead is right, the rest of the story should follow fairly easily. This is true, although, as we saw in the story about the landslide, sometimes there may be several factors which appear to be of almost equal importance.

The second paragraph is usually, a sub-lead or continuation of the lead, and should provide the remaining essential information that cannot fit into the first paragraph. Each succeeding paragraph is used to widen and deepen the reader's understanding of the news, delivering the rest of the facts in diminishing order of importance, the so-called 'inverted pyramid' format.

Inverted Pyramid

Many journalists write their stories in the 'inverted pyramid' form – a structure in which the most important news is placed at the top, followed by the next most important and ending with the least important. That form was invented in the days when mass circulation newspapers in London and New York were new and competition between them fierce. Newspapers published several editions every day, each one striving to be more newsy, more up to the minute than the last. To make room for new stories, older material, no matter how important, had to be cut. To make that easier, the inverted pyramid was invented, allowing sub-editors to cut stories from the bottom without having to think about destroying the basic sense of the story.

If facts had to be sacrificed, cutting from the bottom meant that the least important facts were lost. Apart from those considerations, the inverted pyramid very often becomes a straitjacket or a Procrustean bed which forces writers into unnatural ways of telling a story. Surprise and natural drama are often the major casualties of the inverted pyramid.

If the inverted pyramid helps you to organise your story, use it. If it doesn't, try to tell the story in your own way.

As we saw in the last chapter, there are many ways of beginning a story. Logically, each beginning should suggest how you tell the rest of the story. The fundamental principle is that you must remember that you are telling a story, a narrative, so that even if your approach requires that some facts are presented out of their natural order, the order you have put them into should have its own logic, its own order and create its own narrative.

Using Other People's Words

Reporters, in their zeal to include every available fact, too often forget the development of narrative, the connections between cause and effect. The reader has to work out the connections. Many don't bother.

Quotes may be used at any time appropriate and should be used as early as possible, and as often as is useful. Use quotes to present or illustrate, if possible, the differing points of view of those directly involved. Explaining what people believe is one thing. Using their actual language often provides something more colourful and more valuable – an insight into the way they think.

Quotes provide windows into a story: they give the point of view of someone inside the story, a point of view distinct from that of the reporter.

Quotes should be spread throughout the story, if possible, to illustrate different facets, lighten the texture of the story and to make it communicate its message as efficiently and as painlessly as possible.

Transitions

The writer should try to move the story naturally from point to point, using transitional phrases or sentences to help the reader to understand where he is in the story and to help him understand the relationship between various elements of a story.

Sudden shifts of context, place, mood or viewpoint tend to make the story less coherent and unsettle the reader. Your reader (or listener) may not be able to say what's wrong with your story, but he will sense that something is wrong when you jump without warning to a new idea. Instead of concentrating on the story the reader or listener then starts trying to figure out what is wrong. He thinks he must have missed something, but what? Some people simply give up; they don't know what is wrong and they won't bother to find out.

Transitions are like hinges, connecting one idea, one paragraph, one development or action to the next. A well made transition links one set of facts, one scenario, to the next, making possible logical, smooth progress in the action and development of the story.

Good transitions help the reader to understand the connection between all the factors in the story and help make sense of it. In radio, disc jockeys have a similar mechanism, called a *segue* (pronounced seg-wey) from the Spanish word *seguir* meaning "to follow". As one tune ends, its volume is gently faded down while the next is gently faded up, very gradually, so that there is (ideally) a seamless transition between the tunes – luring the listener into following the programme without having to make conscious decisions.

Editorialising

News reporters should never include their opinions as if the opinions were part of the facts.

The reporter is privileged to be present on behalf of the public. That is what he is paid for, and the public is entitled to the plain and unvarnished facts. The position of reporter, no matter how highly paid, does not guarantee wisdom or confer omniscience. When the reporter meddles in the story, he is simply making it more difficult for the audience to get at the truth.

Journalists and those who perform journalistic functions should therefore avoid sticking on labels and using adjectives reflecting their opinions or their prejudices. They should not inject themselves into stories.

Whenever someone is quoted, there should be instant identification of that person. The public is getting wise to the fact that some journalists quote 'unnamed observers' simply as a mechanism for putting their own opinions (or those of another reporter) into the mouth of some unidentifiable but presumably 'reliable' source. Since these sources are often alleged to be 'diplomats', it is useful to remember Lord Palmerston's description of diplomats – "men paid to go abroad to lie for their country".

When stories include favourable or unfavourable statements about someone or something, these statements must always be attributed to an identifiable person, and should not appear to be the reporter's opinion or the opinion of a (possibly) invented person, slipped into the narrative to mislead the audience. You lose credibility that way. You may also lose money, if the statement turns out to be libellous.

For an example of editorialising, look at this lead:

> Nurses in our country might soon find themselves in the unfortunate position where they will not be entitled to be granted offers to work in the United States of America.

Why is the position unfortunate? Who says it is unfortunate? Obviously – the reporter. It is never explained what gives him this authority. (There are other faults with the paragraph, for instance: ". . . will not be entitled to be granted offers to work" is an incredibly clumsy construction.)

Sometimes, as in the paragraph above and in the next one, entire sentences are purely the expression of the reporter's (or editor's) opinion, without any other basis.

> Serb police chief Simo Drijaca gloated that none of the 9,000 Muslims he says applied to leave wanted to remain in the mayhem that is Bosnia.

That sentence makes it perfectly clear that the reporter has taken sides ('Drijaca gloated'), and that makes his objectivity suspect. The reporter is entitled to his opinion, but opinion should be clearly identified as opinion and not passed off as fact.

If the reporter sincerely believes Drijaca was gloating, that behaviour should be displayed to the reader. It will take a few more words, but since the fact is important, the extra words are worth it. Perhaps a quote from Drijaca would have made his 'gloating' obvious. "The mayhem that is Bosnia" suggests that Drijaca is also gloating about the mayhem. He may have been, but the reader has to take the reporter's word. Even if you have no reason to distrust the reporter, you don't have to accept his judgment. Of course, reporters are not eunuchs or ciphers. They all have their own perfectly legitimate points of view. What is not legitimate is to disguise bias as fact.

Summing up

News stories should be crisply written, clear and factual. They should be easy to understand. News stories should be as complete as possible.

The narrative should flow easily, tempting readers to read the entire account. The message should be simple, clear, unambiguous and leave no unanswered questions.

This brings up an important caveat for self-publishers: even the best editor should avoid attempting to be the final editor of his or her own work. The writer is usually too close to the work to see mistakes or infelicities that may jump out at someone else.

When I edited a one-man weekly, I always asked my part-time proofreader or one of the linotype operators to read my column before I gave it out for typesetting. In those ancient times, linotype operators and proofreaders once or twice picked up errors of style when they were doing what they were paid to do. Many errors passed them unnoticed because many did their jobs on auto-pilot. But the same people, asked to read a piece of copy before it got to them 'officially', would often spot stylistic errors which they would miss when they were doing their assigned jobs of typesetting or proofreading.

8

Making Words Work for You

Many scientists and other experts find it difficult to explain themselves to people who want to learn from them, because the experts use language which is too difficult for ordinary people to understand. Some of this is because every trade, art or profession has its own jargon or private language. These dialects developed in the interest of more precise communication within professions and trades.

When a printer talks about a 'pica' or a doctor about a 'neoplasm', their professional colleagues instantly understand exactly what they mean, but a printer's 'layout' is very different from a surgeon's 'layout' and both are different from an architect's 'layout'. If they are to understand each other they need to speak the language they all share – the language of ordinary people.

Television dramas about hospitals, courtrooms and some other activities have made some jargon part of some laymen's ordinary speech and have given some doctors and lawyers a false sense of how easy it is to understand them.

Some scientists and other experts simply do not understand that their daily language and the way they think may not be at all familiar to ordinary people.

Dealing With Jargon

I have heard the word 'secular' used on a television talk show and wondered how many of the audience would understand the reference, since it was not clear from the context. The person speaking was talking about floods, and secular referred to the probable frequency of very large floods over very long periods of time. It is likely that most of his listeners were completely mystified by the reference, since most people understand secular to mean 'not of the church'.

One of the reasons people use jargon is conceit, to show off the fact that they belong to a special breed or elite, with its own mysteries, its own arcana. I imagine that when they speak they really hope to influence their audiences, not to confound them and mystify them. If they must use jargon it should be used in contexts that explain or suggest the meaning. Otherwise, most of what they say will not even be heard by their intended audiences.

The human mind is capable of many things, but not many minds work like tape recorders. Most people can remember clearly the last five or six words they hear and can readily grasp meanings contained in sentences of twenty words or less. Most people hearing an unfamiliar word will switch off for a few milliseconds to see if they understand it and then switch back into listening. If they don't understand the word they may continue listening. If too many 'foreign' words come at them, most people will cease to bother and will simply switch to something they do understand. Even if they don't switch off permanently, they lose almost all the information transmitted while they were busy trying to understand one strange word.

In many former British colonies, civil service officials have developed styles of writing that seem designed to avoid making sense, while giving an impression of significance and profundity.

The style is characterised by indirection – an inability to get to the point; by the use of the passive voice – people do not act, they are acted upon, and things happen, as it were, without anyone in particular being responsible. There is a deliberate use of euphemism, so that things never seem as bad as they are and, generally, no one can be blamed for anything.

Take this example:

> The Department is fully aware of the fact that maintenance of growth and development, and the motivation of staff to give superior performance, thereby improving the quality of life and productivity, must be of paramount importance. In response to the foregoing, a wide range of internal and external courses is

available. Seminars and relevant internal conferences are also accessed. A total of several hundred persons were trained during the last year, consuming nearly three thousand five hundred hours.

Somebody actually inserted that paragraph into a speech which was to have been delivered by a minister of government. Fortunately, the minister read this speech before attempting to deliver it. The paragraph is confused and confusing, besides being deadly dull and in real terms, meaningless. The information it contains could easily be presented in a more interesting way – in words and constructions that more people would understand. The whole statement can easily be made less confusing:

> The Department expects its staff to perform at the highest level, and has therefore introduced a wide range of training courses and seminars. These courses are offered both inside the institutions and outside. Several hundred staff members took these courses last year.

The original included unnecessary information (3,500 hours) and was ungrammatical, perhaps because it is difficult for such long sentences to make grammatical sense. The language was pompous, striving too hard to make a good impression. The meaning was obscured by subordinate clauses and too many commas.

Let's look at the first sentence again:

> The Department is fully aware of the fact . . .

Of what fact is the department fully aware?

Is it that "maintenance of growth and development" must be of paramount importance and if so, whose growth and development?

Or, is it that "improving the quality of life and productivity" is of paramount importance. Whose quality of life and whose productivity? And how, I ask you, does anyone 'consume' hours?

When you analyse some writing you find that much of it does not make sense.

There are too many famous examples of official jargon for me to bore you with another. Fortunately, as people become more conscious of their rights, they seem also less tolerant of humbug and indecision. Younger people seem to be more conscious of their public responsibilities and the need to be more accountable (and comprehensible) to the public they serve. Obviously, responsible attitudes conflict with irresponsible language. But ever so often one can still find writing of the most numbing obtuseness.

Several decades ago the British Civil Service commissioned Sir Ernest Gowers (a senior civil servant and a classical scholar) to write a primer to guide civil servants in writing. The result was *Plain Words* – a masterly guide to the art of simple, clear and effective writing. I recommend it for your bookshelf.[1]

Jargon is a subtle enemy. It creeps into ordinary speech and soon becomes an established weed, self-seeding, difficult to eradicate.

'At this point in time' is one of those expressions that seem to fill the need formerly satisfied by "Er." and "Aahm" and other verbal spacers used to fill otherwise dead air and allow the speaker time to think. I do not understand why 'at this point in time' is considered superior to 'now' – unless of course, 'now' is too short, too crude, too brutal.

A medical doctor is perfectly within his rights to use words such as neoplasm and haematoma in any context where absolute precision matters, such as in a courtroom. But the ordinary person will more easily understand tumour than neoplasm and blood clot makes more sense than haematoma. If a reporter must quote such language, he should make the meaning clear so that ordinary people will understand what is being discussed.

The Cataclysm Crowd

Many people who deal professionally in natural disasters fall into the habit of using arcane expressions of their trade in ordinary conversation such as a 'Force Three hurricane' or '4.5 on the Richter scale'. There is nothing wrong with using these expressions, they help to educate people. But if we start from the premise that people should understand what we say, it makes much more sense if people were also told that the hurricane was exceptionally violent – the sort that is not expected more than once in a hundred years – or that the earthquake was mild and nothing to be worried about. People who have just experienced a Force Three hurricane would not need to be told that it was violent nor would people in a 4.5 earthquake think they had survived a disaster. But their friends and relatives a few hundred miles away might be lulled either into a false sense of security by the one or panicked by the other.

Since most people don't know how disasters are measured, Force

Three seems comfortably distant from Force Ten – if one presumes that most things are measured on a one to ten scale – except that the hurricane scale stops at five. Most people who have heard of the Richter scale imagine that it too stops at ten. It doesn't. No earthquake has yet been measured at over Richter 9, but the scale is open-ended. There is no top number, and that is a really terrifying thought.

Since most people have only the vaguest idea what these measurements mean, it is silly to use them without some sort of qualifier. When you do explain, you are informing and educating at the same time.

Weather forecasters in my part of the world use the expression 'partly cloudy' to cover a multitude of conditions. What do they mean? Unless one lives in the Sahara Desert or some similar area, skies are usually partly cloudy. A sky with 10 percent cloud cover is 'partly cloudy', as is one with 75 percent cloud cover. 'Mostly cloudy' would describe one condition, 'some light cloud' might describe the other. 'Partly cloudy' describes nothing.

Making Words Work For You

Engineers, particularly (in my opinion) hydraulic engineers, enjoy a really abstruse jargon. They tend to speak in terms of 'acre-feet' and 'cubic kilometres' which are excellent for boggling minds but meaningless to most people.

It is difficult to visualise enormous amounts of anything, especially water, but it is certainly easier to understand if one said that the water would cover all the houses in the whole of the next parish or county. It would become even easier to understand if related or compared to the size of something most people are likely to be aware of, for instance, the local dam, or harbour or whatever comparable large body of water is handy. Most people have a good idea of how big a sports arena is, for instance, or how high their nearest mountain is, and they know the height of the average coconut tree. Comparing strange things to known things makes them easier to grasp. I have never forgotten a newspaper description of the first manned Russian spacecraft as "about the size and weight of a Volkswagen". (In those days, Volkswagens came in one size only.) When I read that, I had a more realistic understanding of what the spacecraft was like. Described in kilos and cubic meters, it remained an abstraction.

Many reporters feel a little nervous asking a scientist to explain particular concepts. They think that they are expected to be aware of the nature of quasars, mesons and other esoteric concepts which now zoom about in the world of television and magazines. The Big Bang theory is one of these concepts which everybody takes for granted but most of us do not even begin to understand. If concepts such as these are vital to the understanding of what you are writing, don't be shy about asking the experts to explain. If you are too bashful to admit your own ignorance, ask for a 'definition I can use so that ordinary people can understand'.

Shyness is one of the more serious bars to knowledge. Some of us are either too shy or too proud to admit that there is something we don't know. The longer we live, the more we realise how much we do not know. I therefore have absolutely no qualms and no shame in asking someone who knows, to explain to me something that I don't know or that I am not sure about.

If you want to make sense, you must understand what you are saying. If you use words and concepts you don't really understand, you are fooling yourself, committing a fraud upon your readers and misleading them, perhaps dangerously.

Your Own Style

Aspiring writers always seem to be concerned about developing a 'style'. They may admire some famous writer or broadcaster and want to know how to develop a 'style like that'.

Everybody already has his or her own style. You already have your own style. It may not be the most elegant or eloquent style, but it's yours. Style is an expression of the way each person thinks, an expression of habits and influences of which the 'stylist' himself may be unconscious. One's style is as individual as one's fingerprints. You may need to develop or improve your style, you don't need to invent one.

The best way to develop and improve your own style is to try to strip away everything that you know may have been copied from someone else. Try to express yourself as plainly and simply as possible in your own voice and in your own words. Your language is your life, the truest expression of your personality, your culture, your self, the essential you.

People are very quick to see affectation and pretension. Some see those things even where they don't exist. If people begin to worry about elements of your style they will soon stop listening to you and start listening to your style instead. Then, what you want to say gets lost in the 'style'.

It's easy, of course, to advise someone else to 'be yourself'. It is not so easy for you to decide what 'being yourself' means. The surest way to be yourself is to stop worrying about yourself, to think instead about the people you meet and about their concerns.

In journalism, and probably in most things, this means simply taking care to make yourself as plain and clear as possible and to avoid the temptation to impress other people with your knowledge or importance. The poet John Dryden said: "The chief aim of the writer is to be understood". That was true long before Dryden (1631–1700 AD) and will continue to be true as long as there are writers and people to read or listen to them.

When you express yourself plainly and simply you are showing your respect for the person with whom you are trying to communicate. It shows that you want to communicate – that you wish to be part of the other person's world and that you want him to be part of yours. It shows that you are at ease with yourself and helps to put the other person at ease with you.

I go into this detail because I do not believe that it is possible to separate personal style from substance. Your style is a part of you, like your skin, not something you put on and take off, like a uniform.

If people are to believe what you say or write, they must believe that you are worth their time and attention; you must be seen to be sincere, trustworthy, truthful. They must believe, as they say in the courts, that you are a witness of truth, and not affected, dishonest or devious.

Clarity In Language

In successful communication, clarity is the most important factor. We want to be understood, so we try to make our meaning as clear and straightforward as possible. My old English teacher spoke of 'limpid' language, as clear and pure as a mountain spring. Our messages must not be misunderstood. Our words should mean exactly what we intend them to mean.

If words are to mean what we want them to mean, we need to be precise. We do not generalise about facts. We do not, for instance, say 'head of state' when we mean 'head of government'; we don't say 'several dozens' when we mean 'fifty' and we don't say 'recently' when we mean 'last week'.

- If we want to be clearly understood, we must speak or write clearly.
- Make sure what you say is what you mean.
- Do not say: 'did not preclude the possibility that' when you mean 'it was possible'.
- Verbiage is the enemy of clarity. Many of us, trying to impress, use three or four words where one would do.
- Use simple language and avoid complicated words if possible. For instance, do not use 'prerequisite' when what you mean to say is that certain things have to be done before certain other things may be done. Using a word like prerequisite can save several words, but it is a difficult word for many people. In this case, brevity may be sacrificed for clarity. Prerequisite should in any case, be avoided in ordinary writing. It looks and sounds too much like perquisite, and, depending on how it is used, can sometimes mean almost the same thing. Compare

    He demanded, as a prerequisite, that the company agree to pay his rent.

with

    He demanded, as a perquisite, that the company pay his rent.

We should always try to avoid using words that do not immediately explain themselves in the context in which they are used.

We need to be at ease not only with ourselves, but with our language. We need to understand the basic rules that govern our language, because these rules also govern our thought. Language is the writer's toolbox and he needs to understand the purpose and functions of his tools – words – if he is to use language to its best effect.

Note

1. Gowers, Sir Ernest. 1962. *The Complete Plain Words.* Harmondsworth: Penguin Books.

9

Feature Writing

The major difference between feature writing and news writing is time. Writing straight news generally takes less time than feature writing. In newspaper parlance, features are anything that is not news or advertisements. Crossword puzzles are 'features', as are comic strips, opinion columns, editorials and articles on cooking. For our purposes, feature writing means the writing of news features – features based on news – which attempt to give background and perspective to public issues – to issues in the news or issues that we think should be in the news. They provide a more rounded picture, making the issues easier to understand and explaining their significance in the kind of detail impossible in a news story.

A news feature is sometimes described as 'interpretative reporting', a name which helps to distinguish news features from news stories, commentaries on the news or news analyses.

The backbone of the news feature is reporting. It is reporting of a different kind from ordinary news reporting because, although it is informed by what is new, it is not limited by it.

There is a famous story about three blind men being asked to describe an elephant. Of course, each could 'see' the elephant by using his sense of touch. The one standing in front, next to the elephant's trunk, described the elephant as being like a large snake. The man in the middle described the elephant as being like a wall and the third, near one of the legs, described it as being like a post. In a crude way, the ordinary news report is a little like the perceptions of each of these men: the reporter is restricted in his reach and therefore limited in his perception. The interpretative piece gives him the space to stand back and walk around the elephant, describing it from all angles and making it comprehensible.

The interpretative story allows the reporter to incorporate in his story a variety of issues supplementary and complementary to the main issue, allows the inclusion of many different points of view about the central issue and related issues and allows him to include background information which provides depth to the overall perception and understanding of the issues.

Let us look at a fictitious scenario.

Environmentalist Accused

Mr Richard Roe, prominent manufacturer, formerly on the board of an environmental regulatory agency, known to have political ambitions, is arrested and charged with illegally dumping toxic material. Roe is said to have hired a company that specialises in illegal waste disposal. Roe is well known in high political circles and it is believed that he was planning to run for office quite soon.

His friends are amazed at the news of his alleged offence and his subsequent arrest. To them he was the model of the environmentally correct businessman, and his life the picture of placid suburban perfection. The police say that they had their eye on Roe for a long time, for other reasons. They suspected him of being involved with a gang of cocaine dealers. They say Roe was also plotting his wife's murder because she found out about his criminal associates and about his affair with a hostess from a nightclub. The wife had hired a private detective to shadow him.

The story suggests several possible approaches to a news feature about the crime and its author.

First, the man himself:

Who is he and how did he get that way?

If we wanted to start at the very beginning, we would try to find Roe's parents or people who knew them. What were they like? How did they treat their son? We could seek out Roe's schoolmates and his neighbours.

What was he like as a child? Did he show any signs of delinquency? We would ask his schoolteachers about Roe's record at school, about his behaviour, his interests, both academic and non-academic. Who were Roe's friends and what did they do together? Did he ever get into trouble? What sort of trouble? How was it resolved? We would ask the same questions about his high school and college careers.

We would talk to the police, find out how long they had had an eye on him and what provoked their suspicions. We would try to find out as much as possible about Roe's alleged criminal associates and how he got involved with them. Is any of them willing to talk? What do they think of him?

We would talk to Roe's former colleagues. What were their experiences with him? How did he get to be a member of an environmental regulatory body? How did he behave as a member of that body? Were any of Roe's alleged gangster associates ever his clients or appear before Roe when he was an official? How were they treated? How did Roe's political ambitions manifest themselves? Were Roe's ambitions realistic? What were his politics?

We would try to talk to Roe's former associates. Who are they? Can the police give any information about them and their activities? Were they all perhaps involved in some larger criminal conspiracy?

We would talk to Roe's friends. What did they think of him? Was he secretive or accessible, solitary or gregarious? What were his favourite pastimes? Do they remember any odd behaviours? Is there anything strange in his past, especially his recent past, to suggest that he might have been leading a double life?

We would talk to the private detective hired by Roe's wife. When was he hired and why? What were Mrs Roe's instructions? What gave her reason to suspect her husband? Who was the other woman? Will she talk? We would try to talk to as many people as possible who might be able to give us anything to explain or contradict the 'picture of placid suburban perfection'. Was there anything which did not fit?

Planning the Story

Usually, when one has gathered a great deal of information about somebody or some event, one or two significant facts stick out, like rocks on a playing field. Patterns begin to emerge, because individuals tend to be fairly consistent in their behaviours and to make the same kinds of mistakes over and over.

These prominences and patterns may suggest a particular approach to the story. Perhaps many people feared Roe or distrusted him. Perhaps he was a charming man with a perfect cover for his double life; perhaps most people all along suspected that something was wrong; perhaps, perhaps . . .

Whatever seems most intriguing to us as we survey Mr Roe's fields of activity will probably seem intriguing to other people too. A 'charming scoundrel' may be hackneyed, but if that is what he was, we need to find a way to explain it without resorting to cliché.

The best approach is to let the facts speak for themselves, to attempt no judgment, but to paint an accurate picture which can be interpreted by anyone. The commentary can always come later.

What Kind of Story?

Having decided on the approach, we need to decide how we are going to tell the story.

Will it be largely a narrative, a small history of a life, or will it be more rounded, a profile in depth (if there can be such a thing), a character study, illustrated by the comments and opinions of people who know Roe? Or will it be an impressionistic piece of reportage?

There are almost as many possibilities as there are writers in the world. You need to make up your own mind, based on what you know of your story, your audience and what approach you feel will give you most scope to tell the truth as you discover it.

Having decided on the approach and on the kind of story we wish to tell, the next step is deciding on the structure of the story.

It is a truism forgotten by many writers that stories need to have a beginning, a middle and an end.

A feature should be constructed to give a complete, rounded picture, so that at the end the reader is left satisfied with the information presented. No writer can hope to capture the whole of any person's character, not even his own, but one should be able to condense the essence of the public face of anyone.

If Mr Roe's life in retrospect seemed predestined for disaster, that may provide the mechanism for beginning and ending your story. If, on the other hand, his newly reported behaviour seems absolutely out of character, that provides another way of dealing with the story.

Whatever is decided, you should make sure that the story moves logically from the start to the finish, pausing here and there to examine particular points in greater detail, illuminating particular features with the comments and judgments of those who know the story and the man best.

Moving logically does not mean moving chronologically. If we find patterns, identify each and move on to the next. In any interpretative story, it is a good idea to talk to people who are experts in your subject. In this case, most of them will be the people around Mr Roe. But there may be aspects of his behaviour which may need to be illuminated by psychologists or similar experts. Use them very carefully.

I have used this invented story because it contains features familiar to most of us who follow the news. I have used it because it provides more circumstantial detail than most stories about the environment or similar matters. The subject may be different, but the technical approach is similar.

Finally, a tip. I never start writing a story before I have a pretty good idea of how it will end. That way I can see and try to plan a clear route through the facts from beginning to end. This technique also helps me to get rid of surplus material which may be interesting in its own way, but which may be peripheral to the real story.

10

The Art of the Interview

According to the *New Shorter Oxford English Dictionary,* the first meaning of 'interview' is: "A meeting of people face to face, especially for the purpose of consultation". The second meaning is: "A meeting or conversation between a journalist or radio or television presenter and a person whose views are sought for publication or broadcasting; the published or broadcast result of this". There are other meanings. An interview can also be an examination or an interrogation.

All of us have heard about these different sorts of interviews and most of us have been involved in some of them. Most of us have had to be interviewed for jobs, some of us have been interviewed for a place in school or college and some of us have even been interviewed by the police, because they thought we knew something which they wanted to know. We are continually interviewing each other, exchanging information or ideas; but the sort of interview we are about to discuss is a little more complex than the casual encounter with a friend. We have all seen, heard and read news interviews, some of us have even had microphones thrust into our faces by reporters in search of a 'sound bite'.

Why Interview?

Any story has a better chance of being taken seriously and being believed if people hear it from more than one source. And the more credible – believable or trustworthy – the sources are, the more likely the story is to be taken seriously.

There is no advertising as potent as word of mouth promotion. One American motor manufacturer was very successful with the slogan 'Ask the man who owns one'. Sales managers use this principle when they get stars like Pelé or Michael Jordan to endorse their products. The stars are paid enormous amounts to make these endorsements to convince us that since they wear these shoes or those shirts they must be good, so we will wear them too.

When you try to convince people to change from one kind of behaviour to another, you have the same kind of problem as the manufacturer who wants people to stop buying his competitor's shoe and to start buying his instead. It is expensive to get a star to endorse your product, but you may find it is almost as effective to use the personal testimonies of ordinary people. Interviews can be your word-of-mouth advertising.

People tend to be impressed by the personal experience of 'the man who owns one' – someone like himself who has found a more convenient or profitable way to do something. The testimony of a reformed criminal or drunk is more credible and convincing to a practising criminal or drunk than anyone who hasn't 'been there'.

If You Don't Know, Ask

Many of us are petrified by the thought of being 'interviewed' or of having to 'interview' somebody else. Some of us go all cold and clammy, our heart rates increase and we show all the signs of fright and alarm. Why?

If somebody wants to interview you, it is because he or she wants to know something that you know. If you must interview someone, it is because you want to know something that he or she knows. If you don't know something, the best way to find out about it is to ask.

Most people who know something that you don't are flattered to be asked for information. They feel even more flattered if you ask them for their opinions – what do they think?

Journalists use interviews, introducing personal experience, to make the story more graphic or picturesque. At the scene of a disastrous fire or accident for instance, the camera usually first focuses on the scene and then the reporter introduces the facts before talking to someone who was 'on the spot'.

The interview helps put the audience into the picture.

Similarly, in radio, an interview provides a kind of window into the story. You may not be able to *see* the scene, but the reporter will try to make you *hear* the background sounds first, before he begins his report. He may then introduce quotes from an eyewitness – a new 'voice'. The new voice adds variety and colour to the broadcast anyway, and helps to authenticate the news by providing a first-hand account of what happened. The new voice might also provide an expert explanation of why the event happened, or explain how the event has affected the people involved.

Similarly (but not so obviously), in a newspaper report, an interview changes the pace of the report by introducing a new 'voice', a new point of view. This new 'voice' provides a point of view different from the reporter's. As in television and radio, an interview in print helps to add credibility to the story, helps to illuminate it by personal or expert experience.

Interviews, therefore, are a good way to add depth to reporting the news.

The main functions of interviews are to:
- Illustrate – giving a picture of what happened;
- Illuminate – shedding light on how or why it happened;
- Authenticate – adding confirmation and credibility to the news – the 'I was there' factor.

Who to Interview

A news story may be based entirely on one interview; for instance, the experience of the lone survivor of a plane crash. Another news story may be based on several interviews, giving different sides of the story, exploring different dimensions. A story may be based on documentary research, illustrated by interviews with people who have special experience or knowledge of the facts.

If you are writing about an event, your choices are wide. You can talk to the central actors in the event, to those who are or will be directly affected, to those who were present, or to experts who can explain what happened and theorise about what is likely to happen.

The sole survivor of a plane crash is the expert on what he saw and felt. An aeronautical engineer may be an expert on what caused the crash, an eyewitness may be able to say exactly what she saw and did. Each helps to give depth to the story, to illuminate it in some way. None of them may have the entire story, but together their accounts help other people make sense of events which they themselves have not experienced.

The sort of interview you want, therefore, will depend on what you intend to do with the interview.

Let us suppose that you are doing an opinion piece on the bad effects of garbage dumping. You may want to suggest that:
- garbage is simply a resource in the wrong place and that society can benefit by recycling and reusing the valuable materials now thrown away;
- dumping garbage has bad effects on public health, on public attitudes and on public amenity;
- dumping garbage not only wastes the materials which make up the garbage but becomes even more expensive because the costs of waste disposal are unacceptably high and the side effects dangerous.

Local Experts Add Relevance

You can find documentary evidence to support all of those statements, and you could present a synthesis of all of these facts as a straight piece of reporting.

Your story will be much more interesting, important and relevant if, instead of depending on research material, you consult local experts who could give you the same kind of information, but make it more interesting for your audience by speaking directly about your local problem.

They can tell you, for instance, approximately what quantities of paper or aluminium or plastic containers are collected every day or every week. They can tell you how much this waste collection costs the community and they may be able to estimate how much of the waste can be recycled and how much it would be worth.

Other experts can tell you about whether the wastes damage the soil or the underground water resources or the atmosphere, and what that sort of damage costs.

Other experts can relate these costs to other things, things the community needs but cannot afford at the moment, infant schools, say.

Instead of a dry, theoretical piece which tells people what is good for them, you will instead have a piece which may give your community all the information and argument they really need to convince themselves that doing the right thing really makes sense, may even save them some money and probably make their environment and the children healthier.

The Interview

If you want to get the best of an interview you need to prepare for it.

First, you must have a very clear idea what you want from the interview. What point of view, what information. You should have a good idea before the interview of where it will fit in your story.

An interview is most effective when it is used, not only as an illustration, but as a bridge from one part of the story to the next. Let us say that we are interviewing the person in charge of collecting and disposing of garbage. We will want to know from him what is the present situation. This could be the only reason for the interview. But you may make your story more interesting if you try to find out whether he is satisfied with the present situation, whether he has any ideas for improvement. He may, for instance think that what is being done is the best that can be done and may regard recycling as a pipe dream, one of these fads that come and go.

If that is his position and he says so, he has given you a gate, a transition into the other side of the argument, where the conflict between his ideas and your thesis can be explored in an interview with another expert who thinks that reuse and recycling is the answer and who will explain why.

It helps, therefore, to have some idea of what your subject's opinion is on the issue, in addition to knowing what facts he can tell you about garbage collection and disposal. In preparing for an interview, you need to find out as much as you can about the person you are about to interview.

What is his correct name and title, what is his area of expertise, has he written any books or articles on the subject, what can he tell you about what you want to know?

Introducing Yourself

Let us say you are to do an interview for radio or television.

If you have never previously met the person you are interviewing, your first job is to introduce yourself. This simply means explaining who you are and explaining generally what it is you want to know. If your topic is complex and the interview is likely to be long, it is useful to work out with your subject some basic ground rules for the interview.

If you have enough time, it is useful to sketch out very broadly, the ground you want to cover but without getting into detail.

You need to remember that while you may know quite a lot about the subject of the interview, you are doing the interview to bring information to people who probably don't know very much about the matter. Make sure, then, that you ask questions that will produce answers that make sense to your audience.

Be simple.   Simplicity is the hallmark of the good interviewer. He knows that he is asking questions on behalf of people who may be totally ignorant of the subject, so he tries to introduce the subject as easily as possible.

If, on the other hand, your interview is about a topic which has had wide exposure, you don't need to go over ground which has been covered before, but you do need to make sure that people understand the context of your interview and why you are asking these particular questions.

If you are preparing for a broadcast interview, whether recorded or live, don't get into any detail during the preliminary chat. You risk wasting answers which would be perfect for the 'real' interview. The responses will not come out the same way the second time. Many people spend so much time in preparing an interview that all spontaneity is lost. Often, the best lines are lost and never heard in the broadcast interview.

Some people try to rehearse an interview with their sources, to tell them exactly what questions will be asked in the interview. This usually kills the interview. This is why so many interviews by internationally known 'interviewers' are often so dull. All elements of spontaneity disappear and the interview is flat, stale and tired. If it doesn't seem flat and stale, it will probably be because interviewer and subject are great actors, and that, too, will usually come across and lower the credibility of the interview. If your audience thinks that you are in cahoots with your subject your interview will probably be a waste of time.

The 'pre-interview' is meant to make sure that interviewer and source have a basic understanding of the ground rules for the interview so that neither is likely to be unpleasantly surprised by the other. Take the person into your confidence; explain what you are doing. Explain that you don't want to kill the real interview by rehearsing it. You will be reassuring your subject that you have no malicious intent. It will help you develop rapport.

During the preliminaries, the interviewer can ask the subject whether there is any special aspect of the discussion on which he wishes to make a particular point.

Do not allow the subject to make his point at this juncture. All you need to know is the *area* in which he needs to make the point.

The question is intended simply to give the interviewer (you) a little more insight into the person being interviewed, to give a lead perhaps to areas which you may not have considered, because you did not know your subject well enough. The point is to skim, stimulating your subject to think a bit about things that really concern him and perhaps enable him to find a word or a phrase which will light up the entire interview.

The 'pre-interview' is also meant to put the participants at ease with each other, to reassure each that the other means no harm. I like to tell my sources approximately how long the interview will be, which reassures them that they are not being asked to waste time on Amateur Hour. In recording broadcast interviews, I try to reduce the need for editing to zero.

I once had to interview Archbishop Makarios, President of Cyprus. A few months earlier, the Archbishop had been in his palace when it was surrounded by Turkish soldiers intent on capturing him or, at least, confining him to the palace. He managed to escape, no one knew how and everyone wanted to find out.

During the course of the preliminary chit chat, I had asked the Archbishop questions which made him realise that I knew quite a lot about him and his career when he led the Cyprus independence struggle.

These questions had nothing to do with the question we were to discuss in the interview, which was the prospects of Cyprus regaining control of that part of its national territory that was occupied by the Turks. But the Archbishop was flattered by the fact that I was so genuinely interested in him and quickly became at his ease, making jokes, and soon we were chatting like old friends.

When I asked him how he had got out of his palace and evaded the troops surrounding it, he began by talking about all the famous foreign correspondents who had speculated on his escape, some of whom had even asked him how he did it.

"I am telling it for [the] first time" he said, and explained that he had simply used the phone in the palace kitchen to summon a taxi-driver – a contact from the revolutionary days. He had walked out of the palace and driven to safety in a cab.

It was a great story and what some called a world scoop. I think I got it simply because I had been genuinely interested in him. The recorded interview was 20 minutes and the whole process took less than 45 minutes. Preparation is everything. There is no substitute for homework.

The lesson is that people, even prime ministers and presidents, are human and want to be seen as people and not as objects. If you have a genuine interest in them, they will respond. And that, in my opinion, is the most important thing anyone can ever learn about interviewing.

11

Interviewing Techniques

The second most important thing I learned about interviewing was – Ask the right questions.

Books have been written on interviewing techniques. Something which seems so simple when it is done well is in fact a very important journalistic technique and not easy to master.

It helps if you have a natural healthy curiosity about people and things, and it is essential that you should be *simpatico*, as the Spanish say, likeable and easy to get on with. The real secret is to be genuinely interested in other people, not as a scientist is interested in specimens, but simply to see your subject as another human being, with feelings, anxieties and insecurities, just like you. Many journalists treat 'important people' as if they are made of steel and concrete.

No amount of sympathy is of any real use if you don't ask the right questions. There is an art in asking questions and in knowing what sorts of questions to ask. The answers you get depend on the kinds of questions you ask.

Every culture has certain stock phrases for certain social situations. The English, for instance, use "How do you do? or Howdido?" as a greeting, formal or informal. In Spanish, French and most languages something similar is standard. Most people normally don't expect a real answer to such a question; it's simply a way to open a conversation. Occasionally, however, some bore will take you seriously and give chapter and verse about his latest trials and tribulations, about his car, his dog and his laundry.

It is not the sort of question anyone expects to be asked in an interview, although many of us will have heard so-called interviewers asking their guests questions just as general as 'How do you do?' The usual format is to state the guest's reason for distinction and then to ask: "Tell me about it?"

If you are lucky, your guest may tell you what you really want to know. If you are not lucky, he or she may counter by asking: 'Tell you about what?"

That sort of response can destroy the interviewer and the interview. If you ask silly questions, your guest will probably feel insulted. This person hasn't bothered to find out anything about me, he or she will think. You won't get much of an answer and you may even be insulted. The best that can happen is that the interview will tend to meander, purposelessly and pointlessly, until time enforces a merciful halt.

Questions like "Tell me about it?" are called open questions, because the questioner sets no bounds on the ground he expects his interlocutor to cover.

This may not be much of a problem if the interview is not for broadcast. If it is not for broadcast, you can simply and unprofessionally, cut in and formulate a better question. If the interview is being broadcast, however, you may find yourself carried into areas which you never intended to explore, about which you know nothing and couldn't care less about.

It will be your own fault.

If your interlocutor takes the bit between his teeth and gallops off, you have a real problem. Your problem will be especially acute if he starts to talk about something which is really interesting, but not what you want to talk about. To rein him in and bring him back will make you seem rude and thoughtless to your audience. And they will be right.

Strategy in Interviewing

Interviews need to be planned. That is, the strategy of the interview needs to be planned. We need to know where we are going and in general terms, how we are going to get there.

I tend to outline the major subjects I want to discuss. I write down the subjects. I do not write down questions – that is, most of the time I don't.

Sometimes, of course, there is a question or two that must be asked. I don't write down the question itself, but make a note to make sure I ask it. The formulation will come out of the answers to my other questions.

I try to imagine how the probable answers will fit within the time I have available for the interview. This means that, unless I am interviewing somebody for a personal profile, I try to concentrate on as few lines of questioning as possible – preferably just one – so that when the interview is over I should have a fairly rounded, coherent and self-contained story.

You must know what you want or you will probably get nothing you can use.

Open and Closed Questions

My first questions tend to be fairly open, to give my guest some space in which to define his/her position. With each question thereafter, I try to close the range a little bit more every time, so that at the end of the interview, my last question should round off the interview, tying up loose ends and making my story complete.

Communications experts have defined a whole range of possibilities between completely open questions and completely closed questions. Completely open questions allow the guest space to roam, completely closed questions allow no space at all.

An example of a completely closed question (in law, a 'leading question') is one to which your unfortunate subject can answer only 'Yes' or 'No'.

Even in courtrooms, this sort of questioning is rarely allowed. Unless one is conducting an interview with a shady character – somebody not to be trusted to tell the truth – closed questions tend to shackle the interviewer as well as his 'victim'.

We are trying to get information – not conducting a 'third degree'. If you want information, rather than hoping to score points, questions should be restricted only enough to guide the guest in the direction you wish to go. Anything more restrictive, if prolonged, tends to make guests feel trapped and uncomfortable and they soon stop communicating and begin to answer in monosyllables.

As a general rule open questions tend to be shorter, closed questions tend to be longer.

In closed questions what happens is that the interviewer is trying to build a verbal fence to confine his subject, and so needs to use more words. Open, free range questions tend to be shorter. This is not always true: "What's your name?" is very short, but it is a closed question, because there is normally only one possible answer.

The ideal question is one designed to get exactly the information you want.

A very great lawyer once told me that in court he never asked a question to which he did not know the answer. An interview is not a cross-examination, obviously, but although the situation is different, you should, like a good cross-examiner, have a pretty good idea of the kind of answer you are likely to get to most of the questions you ask.

I was once involved in a potentially explosive situation as another journalist and I began a television interview with the Indian Prime Minister, Indira Gandhi. My colleague asked Mrs Gandhi how long had she been in politics. Mrs Gandhi snapped back: "How long have you been a journalist?" She seemed on the brink of walking out of the studio. To defuse the situation I then asked Mrs Gandhi a closed question which I thought would intrigue her and which demanded an answer. Her whole demeanour changed in a flash. She became interested in talking to someone who obviously knew something about her life. She opened up and we got a great interview.

This was my question, which I asked as casually as I could manage:

"Prime Minister, when you were a small child you were used as a courier to smuggle notes between your father in a British prison and his party outside. It must have been very dangerous. Did you understand what you were doing?"

"Yes. I knew we were in a desperate struggle, a life and death struggle."

"And weren't you ever afraid of what might happen if you were caught?"

"Yes. I knew my father would get into even more trouble, so I made sure I was not caught."

"How did you do that?"

"I simply decided that what I was doing was perfectly proper. Since it was innocent, I was not afraid and since I was not afraid, I was not caught." We had a flying start into the interview, which incidentally, was about then current political questions and not about her personal life.

Closed questions tend to suggest or sometimes even contain, their own answers. In courtrooms, these are called 'leading questions', because they lead the witness in the direction the lawyer wants him to go. They tell the witness what to say. Some people will say what you tell them to say, and it may not be the truth.

Although a closed question worked so well on that occasion, on the whole closed questions tend to provoke hostility. People resent the feeling of being hemmed in, of being restrained. Some people will try to break out of the pen by using a very simple device: they answer a question with a question.

The actress Tallulah Bankhead had a very deep voice. She was once asked by a male reporter: "Miss Bankhead, have you ever been mistaken for a man?"

"No," she said, "Have you?"

Keep your questions short. The longer you make them, the shorter the answer will be. Longer questions may also confuse the person being interviewed as well as your audience. If your subject starts asking the questions, you will know that you are in real trouble. If you ever find yourself in such desperate straits, you could say: "Will you please answer my question?"

(Since it will usually be your fault that there was no real question to answer in the first place, your response may seem outrageous to your subject, and to your audience. And it may turn your subject into a hostile witness, but by this point you have little to lose.)

You then, hopefully, state the real question in question form and not as a statement. If you are lucky, you may get an answer. If you do, keep asking short questions and hope for adequate answers. And pray that the stopwatch is in overdrive.

In my pre-interview with Mrs. Gandhi, I stated the context and then asked the question. I did this because my question was unrelated to what we had come to talk about. But she had been so upset by the other interviewer that she gave every sign of wanting to get up and walk out of the studio. Although my question was totally irrelevant to the real interview, I introduced it to calm the atmosphere, to intrigue Mrs. Gandhi with something unexpected yet familiar to her, to get her interested in being interviewed by someone who knew who she was and seemed to be really interested in her.

The catch is that you cannot pretend to be interested in someone if you are not. So, if you are not easily interested in your subject, don't ask for an interview. It will be painful for both of you.

Another point: unless you are both professionals, two interviewers can create confusion instead of enlightenment. I prefer to do my interviews without assistance. If you have to be part of a team, as in a news conference, the best strategy is to try to follow the general line of questioning before breaking away with that question that you absolutely must ask.

The more smoothly and logically any interview is conducted, the more information is likely to be produced. This goes for news conferences as well as one-on-one interviews.

People do not like being jumped from one subject to another. Most people think allusively – one thing leads to another – and you will see why the best interview is really very much like a normal conversation between two intelligent people, except that in a conversation, there is usually an exchange of views. An interview may sound like an exchange of views, but it really is quite different. If you understand that, you should have no problem with interviews.

Microphone Etiquette

If you happen to require a microphone for your interview, please do not treat the mike like a table-tennis racket. Even supposedly experienced interviewers can be seen swinging the mike back and forth, between themselves and the person being interviewed. There are several things wrong with this.

- If the interview is for radio, the change in the position of the mike can make the background sound vary so much that listeners may think the interviewer and the subject are in two different locations. As far as sound is concerned, they are. Depending on what sort of background noise each of you is facing, the ping-pong technique can lead listeners to believe that the interview is a fake, spliced together by a sound technician out of two different recordings.

- Moving the mike makes some interview subjects nervous and less able to talk freely. They begin to concentrate on the mike rather than on your questions. And some interviewers forget where the mike is and move it toward the witness when it should be nearer to the interviewer and vice versa. Life can become very difficult.

The correct way, like so many other good things, is much simpler. If possible, stand or sit beside the person being interviewed. This reduces the confrontation level for a start.

Keep the mike in one place, about equidistant between the two of you, checking that the sound level for each of you is right. With the mike steady in one place, both interviewer and interviewee can forget about it and get on with the interview. The comfort level goes up several notches immediately.

These rules apply for all interviews, whether for broadcast or not.

Summing Up

My rules for interviewing are simple:
- Be prepared.
- Make yourself really familiar with your subject's background.
- Make yourself really familiar with the topic you are to discuss.
- Have a strategic plan.
- Know what it is you want to have explained. What is the major issue in your mind?
- Prepare a general line of questioning, not specific questions. Be ready for emergencies.
- If you know your subject, you will be prepared for all but the most drastic accidents.
- Be economical with your questions.
- Don't fence your subject in with words; allow him/her freedom.
- The more words you use, the less your subject will need to.
- Ask real questions that demand answers. Do not make statements.

You may try to establish the context of a question, but it must be clear that this is what you are trying to do and not making a statement. Statements will be contradicted or assented to, but 'yes' or 'no' is not the kind of answer you normally want.

What Is Libel?

Journalism is about facts and commentary on facts. When that most dreaded of sanctions – a writ alleging libel – is applied against a journalist, it is facts that will save him or her. But anyone who writes for publication can be sued for libel, and, as you will find out, publication does not necessarily mean spreading the word to thousands. Simply transmitting the information to a third person is dangerous.

If what you say is in a permanent form it may be libellous. Permanent publication includes a wide range of activities: a newspaper or magazine article, a radio or television broadcast, a theatre skit or play, even a public speech or a letter or email on the Internet. If it reflects dishonourably on someone, you may be sued for libel. And if you lose the case, the costs can be horrendous.

What is libel? To understand, let us go back in time.

Property Rights

Laws develop out of custom and usage, and most human societies have generally considered that people are entitled to enjoy a peaceful life and to deserve protection against gratuitous attacks on their persons and their property. A man's reputation was considered part of his property. Every person has the right to earn the goodwill and respect of his fellow men. If someone publishes a statement attacking his reputation or character, such an attack is regarded as an invasion of his legal right. Attacks against a person's reputation have for a long time been regulated by laws against defamation.

The first prohibitions against defamation applied to the defamation of an artisan's productions or his trade goods (his professional reputation) and the imputation of immoral behaviour against women. These are still the most serious libels.

In the English Common Law, libel was originally not simply a civil offence, but a crime against the society itself, like murder, rape, arson or theft. In earlier times therefore, persons convicted of libel were subject to legal penalties: imprisonment or fines. In the time of the English king, Alfred the Great, eleven hundred years ago, slanderers were liable to have their tongues cut out. Later, lesser slanderers – such as 'common scolds' – were subject to the 'ducking stool' or to be locked into pillories where their neighbours could throw rotten eggs and other obnoxious stuff at them.

More recently, people began to feel that there should be a difference between serious libels and less serious libels.

Defamation was therefore re-defined into several classes:
- Blasphemy was the defamation of God
- Sedition was defamation of the State
- Slander was defamation by words spoken
- Libel was the publication of a slander in permanent form – by writing, dramatisation, by drawing, gesture or other non-verbal means.

Publication Essential

In recent years, slanders broadcast by radio or television have been added to the class of permanent defamation, that is, libel.

In some areas, words spoken over a loudspeaker are also considered libel although the words may never have been written down. Even a wordless skit, on television or the stage, for instance, may be libellous. Libel, then, is a slander generally disseminated so that it may be seen, heard or otherwise become known by people at a distance from the author. It is defamation with *intent to do maximum damage*, not just 'bad talk'.

Publication is an essential ingredient of libel.

Anybody can write you and malign you to his heart's content, but if he does not publish his letter to some third person, you cannot usually sue him for libel.

If however, he writes somebody else a letter in which he maligns you, then he has libelled you, and if you can get hold of the letter, you can sue. If you get an offensive letter about yourself from someone you may, of course, depending on the laws of your country, be able to sue him for harassment or blackmail or something else, but usually not for libel.

There are two exceptions to the publication rule: one is if the libel is a criminal libel; the other is if the libel is so gross that it constitutes an unreasonable provocation; that is, it is so scurrilous that the recipient is likely to be provoked to antisocial action.

Otherwise, the allegedly libellous statement must have been published by the author to some person other than the plaintiff. Generally, as long as the author of a libel against you ensures that no third person sees the libel between his writing and your receipt of the libel, the libel is not actionable, that is, although you may have been defamed, you cannot sue.

If the writer transmits a libel by telegram or postcard, or, like Lord Alfred Douglas, by leaving an offensive message in a public place – he has published the libel, and may be sued. If the writer dictates the libel to his own secretary, he has published the libel.

A telegram or postcard may both be read by third persons and it is technically irrelevant whether any third person has actually read the libel. If, in the ordinary course of things, some third person (such as the addressee's secretary) could possibly have read the note, it is actionable.

If the person who is defamed is the one who publishes the libel, by showing it to someone else, he cannot sue unless the person to whom he shows the libel goes and repeats it to someone else. He can then sue the person who repeated the libel, if he feels he has to do that. He cannot sue the author of the libel, because he was not the publisher.

Criminal Libel

Libel is divided into two classes: criminal and civil libel.

Criminal libel, broadly defined, is a very serious libel which the state considers requires its intervention: for instance, a libel which unjustly imputes criminal conduct or a criminal character to a person.

A slander – a spoken libel – may also be defined as criminal libel if it is likely to lead to a breach of the peace or if it is considered to be seditious. Publications which may lead to a breach of the peace may also be considered criminal libel.

If someone defames a beloved national hero, for instance, the dead man obviously cannot sue, nor can his estate or relatives. But since defaming a national hero may very likely lead to a breach of the peace, such defamation may be considered a criminal libel in countries with a British colonial history. In some other countries, defaming a national hero may actually constitute a criminal offence.

Anyone accused of criminally libelling someone must not only prove the truth of the statements, but must also be able to prove that the statements were made for the public good. The author of the words must be able to prove that the facts he has alleged are so important that it was his public duty to make them generally known. If he cannot prove that his statement needed to be made for the public good, the statement, even if true, may be a criminal libel. This is the one place where the old adage, 'the greater the truth, the greater the libel' may apply.

In former British colonies the Attorney General must approve the launching of any action for criminal libel.

Civil Libel

Civil libel is any publication which tends to deprive a living person unjustly of his reputation, or which might expose him to hatred, ridicule or contempt, or would tend to lower him in the estimation of right-thinking people, or could cause him to be shunned or avoided by his neighbours.

A shorter and more succinct definition is that civil libel is the publication of a false statement or statements about a person to his discredit.

As in cases of murder, rape, arson and other serious crimes, the law provides that a jury of ordinary people shall judge the guilt or innocence of those charged with libel unless the parties agree to dispense with a jury.

Any living person may bring an action for libel; criminals in prison, children and people certified insane may bring actions or have actions brought on their behalf, to protect their reputations. Dead people, however, cannot be libelled and a libel action dies with the plaintiff or the defendant.

In *slander*, except for four exceptions classified as *slander per se*, the plaintiff must prove that he has suffered loss. This is unlike libel, where the law presumes the plaintiff has suffered loss and he doesn't have to prove actual damage.

Once the plaintiff has established that he has been defamed by the libel – that his character or reputation has been called into question – the law *presumes* that the damaging publication is false and since it is false, it is *presumed* to be malicious. The person accused of libel must either prove that what he said was true *and* that he was justified in saying it, or he must provide some other reason to exculpate himself. The plaintiff, that is, the aggrieved person, must prove simply that:

- the words (or other matter complained of) were defamatory;
- the defamatory reference was understood to refer to him/her; and that
- they were published by the defendant to some third party.

In a libel action, the first task of the court is for the judge to decide whether the material complained of was capable of being defamatory in the natural and ordinary meaning of the words.

If the judge finds that the words were capable of a defamatory meaning, it then becomes a matter for the jury to decide whether the words were in fact defamatory. The court may also be asked to decide whether the words may suggest additional things outside their ordinary meaning (*innuendo*) and are therefore libellous.

The plaintiff who relies on innuendo may call witnesses to say that they understand the words complained of to mean more than is apparent from their 'natural and ordinary meaning'.

Defences In Libel

The main defences in libel are:
- Justification
- Fair comment
- Privilege and qualified privilege
- Innocent publication

Justification

The facts alleged are true and could have been proved to be true at the time of the publication. The defendant cannot rely on facts that become known after the publication of the libel. Otherwise, truth is an absolute defence to a charge of civil libel.

Fair Comment

The defendant must establish all of the following:
- the words are obviously commentary and not allegations of fact;
- they must be based on facts which are true or privileged;
- the comment must be fair; and
- the commentator must not have been inspired by malice or some other improper motive.

Privilege

A publisher may claim that a defamatory publication is protected, either by absolute or by qualified privilege. Statements made on certain public occasions are immune to civil proceedings – absolutely privileged – even if they are untrue and damaging. Statements made in parliamentary proceedings or in judicial proceedings reported fairly, accurately and in a timely fashion by the media are all absolutely privileged, no matter how untrue or damaging they may be to someone.

In parliamentary *proceedings,* only the original libel is protected by absolute privilege.

Press reports of a Parliamentary incident or words are protected only by qualified privilege. The defence of qualified privilege fails if the plaintiff can prove that the defendant was motivated by malice. While a reporter may escape liability for reporting a false and malicious accusation made in parliament, a commentator may have to pay damages if he treats the allegations as fact when he knows or should know that they are false.

Qualified Privilege

Qualified privilege also applies to statements made where there is a social, legal or moral duty or interest in communicating relevant information; for instance, an employer's reference or a reference from your professor.

Fair and accurate reports of parliamentary proceedings and fair and accurate reports of judicial proceedings are also, as we said, protected by qualified privilege.

In order to be fair, court reports should be sufficiently balanced that the reader or listener is not likely to be misled. Absolute privilege will be lost if the report of the case is one-sided or lacks impartiality.

In order to be accurate, court reports should contain no significant misreporting of the proceedings. The report should generally reflect what happened in court without any extraneous matter. Reports can only attract privilege if they are about cases heard in open court. Reports of *in camera* proceedings are not protected.

Privilege does not apply to the publication of any matter prohibited by law, such as indecent matter, nor to any matter which is not of public concern and the publication of which is not for the public benefit. In the United States it has been held that criticisms of public servants in the execution of their offices cannot be libellous unless actual malice can be proved on the part of the defendant.

In the case of the New York Times Company v. Sullivan (1964), the US Supreme Court held that the First Amendment to the US Constitution which guarantees freedom of speech protects critics of public officials, even when they make charges that are false. In 1964, L.B. Sullivan, an elected official who supervised the police department of the city of Montgomery, Alabama, sued the *New York Times* for libel, alleging that he had been libelled in an advertisement published by third parties in the newspaper.

Although Sullivan was not mentioned by name, the advertisement alleged serious police misbehaviour at many levels. Sullivan felt that these allegations, some of them false, impugned his own integrity and reputation. The Alabama court awarded him $500,000 damages.

The US Supreme Court overturned the jury's award, holding that the First and Fourteenth Amendments prevented public officials recovering damages for defamatory falsehoods relating to their official conduct unless it could be proved that a statement was made with "actual malice", that is, "with knowledge that it was false or with reckless disregard of whether it was false or not."

Courts in other countries have not seemed prepared to follow the decision in the Sullivan case.

Innocent Publication

Here the defendant must prove that the words were published innocently in relation to the plaintiff; and that the publisher made a suitable offer of amends, including a suitable correction and an adequate apology and that the offer was refused by the plaintiff, but was still open up to the time of trial.

Other Defences

'Leave and licence' is the defence offered by a defendant who can prove that the plaintiff clearly consented to the publication of the libel, as in the case of a criminal who has sold his life story to a newspaper and then sues the paper for publishing facts damaging to his case.

Repeating a Libel

Reporters are always warned to be on the alert against publishing libels originated by other people. Informants (sources) may be misinformed or mischievous, and it is not a defence to a charge of libel that someone else said it first. *Everyone* who repeats a libel can be sued.

In a newspaper libel, the plaintiff can sue the reporter, the editor, the proprietor, the printer, the distributor and the newspaper vendor. A talk show host who allows someone to phone in to libel a third person is as guilty as the author of the libel, even if he does not know the identity of the author of the libel.

Identification

It is extremely important that people in news reports are properly identified. There have been cases where wrongdoers were not properly identified by a newspaper, and people with the same name have successfully sued for libel because they proved to the satisfaction of the court that the report was taken by their neighbours to be referring to them. The statement does not have to be damaging in itself. Mr John Smith and Miss Jane Brown may be well known partners in one locality. But another John Smith may sue if it is not clear that he is not the person alleged to be consorting with Miss Brown. Accidental libels can be just as expensive as other libels.

Damages

Damages for libel may be awarded under three headings:
- *Compensatory* damages comprising 'special damages' for actual pecuniary loss and 'general damages' for the plaintiff's distress and injury to his reputation.
- *Aggravated* damages are awarded when the defendant's behaviour

after the libel has somehow added to the injury or suffering of the plaintiff.

- *Exemplary* damages are awarded where the jury, in addition to compensating the victim, wishes to punish the defendant for his behaviour, as when the libel is repeated, or is particularly reckless or nasty.

There are also so-called *contemptuous* damages, when a jury awards the plaintiff such small damages that it is clear that they believe either that his reputation was worth very little or that the libel was a technical libel only and that bringing it before the court was a waste of time – a 'frivolous' or 'vexatious' prosecution.

Sometimes a defendant may admit and apologise for a libel and pay into court whatever money he estimates would compensate for the actual loss suffered by the plaintiff. The plaintiff will know that the money is paid into court, but the jury will not. If the plaintiff pursues the case after this, he may be at serious risk. If the jury's award is not more than the sum paid in, the plaintiff will usually have to pay the full costs of the trial, since the trial was a waste of time.

If it is the plaintiff who loses in a libel case, he usually has to pay the entire costs of the case, that is, his own costs and the costs of the defendant and his legal advisers. In a libel case, these costs may be extremely high. If the defendant loses the case, he has to pay damages as well as the plaintiff's legal costs.

Chronicle of an Abused Planet

Human beings share the Earth with millions of other living species, all entirely dependent on the planet and its resources for existence. As far as we know, we inhabit the only spot in the vast universe on which we can all survive. Although we have not always treated the Earth prudently, it is only recently that humanity itself has begun to inflict lasting, perhaps fatal, damage to the planet's life systems.

For hundreds of thousands of years, human beings were content to coexist with other natural systems, subsisting more or less peacefully with other species, although always the predator at the top of the food chain.

Lately we have become greedy and wasteful feeders, consuming more than our share and destroying more than we consume.

Over hundreds of thousands of years, humans developed and refined farming, hunting and other economic activities, developed higher yielding strains of food crops and bred improved strains of cattle and other domestic animals. They made their own clothing, developed more sophisticated ways of preparing their food and began to enjoy the leisure that their industry had brought them.

By today's standards their communities were small and produced little waste. What they left behind was soon consumed by natural agents – other animals, termites, bacteria and other organisms.

Mistakes, Ancient and Modern

Environmental degradation is not new. Disastrous environmental degradation is not new either. Scientists believe that some ancient civilisations disappeared largely because of their unwise use of natural resources.

Mohenjo Daro, a region of the lower Indus watershed in modern Pakistan, bears traces of one of the world's most ancient civilisations. Here, nearly three thousand years ago, thrived a sophisticated people who developed elaborate irrigation systems to cultivate thousands of hectares of arable land. They produced great wealth which gave them time for leisure and the enjoyment of life.

The rich farmland is now desert. For many years scientists were unable to explain why such a wealthy and apparently stable civilisation should have disappeared.

After years of scientific research, it now appears that life in Mohenjo Daro became unsustainable about 100 BC because of the very success and sophistication of the irrigation canals. The canals, over the centuries, raised the water table, dissolving chemicals in the subsoil and bringing them to the roots of the crops. Over time, as toxic salts accumulated in the soil, agriculture became increasingly difficult. As food became scarcer, the people became poorer, populations dwindled and eventually either drifted away, starved or became prey to invading forces.

On the other side of the world, in what is now Mexico and Guatemala, man-made environmental disaster may also have destroyed another highly sophisticated civilisation.

The Maya began to flourish in Central America about the same time as the people of Mohenjo Daro. The Maya civilisation was one of great sophistication and complexity. Their calendar measured time from 3114 BC and was at least as accurate as the European Julian calendar which was invented a hundred years after the Spaniards landed in America. The Maya built enormous pyramids and other stone monuments. The stonework was plastered with a cement made mainly with quicklime – made by roasting limestone in very hot fires until the stone disintegrates into powder.

Scientists now believe that the Maya cut so much wood to burn quicklime that they destroyed their forests. This, in turn, caused soils to lose nutrients and wash away. Drought and reduced fertility meant smaller crops, food shortages and, finally, the extinction of the civilisation. The Maya empire lasted a thousand years longer than the people of Mohenjo Daro, dying out about a thousand years ago.

Bruce Dahlin of Howard University, quoted by Stephanie Pain in the *New Scientist*, says history is littered with examples of civilisations which failed to adapt to changing conditions.[1]

They failed to change either because they were too set in their ways or were ruled by an elite too concerned with its own self-interest to change the way they governed.

According to Dahlin "global warming is not the real problem – entrenched modes of adapting to change are." He says we too are relying on technology to get us out of impending environmental crisis. "The developed countries are today's elite class", confident that their systems of trade and economics will prevail, much like the Maya rulers.

Conquest

When Christopher Columbus sailed into the Western hemisphere in 1492, the Maya empires had been gone for nearly six hundred years, but descendants and relatives of the Maya still ruled in the American continent. The Aztecs and the Incas ruled enormous empires. The Aztec empire comprised about 500 cities in Mexico. Europe was stunned by the stories and exhibits brought back by Cortes and those who followed him. The New World seemed to promise unimaginable riches. Its subjugation and exploitation were soon the highest priorities of Europe.

Although most of the people Columbus met were fairly friendly and not very warlike, there were some highly organised societies in Mexico and South America capable of waging war. Nevertheless, within a very short time, and to the surprise of the Europeans themselves, relatively small bands of Europeans had conquered most of the Americas and forced the 'Indians' into slavery. Within a few years native American labour was short and the Europeans began to import labour from Africa.

Although Columbus found thriving populations of 'Indians' in the Caribbean islands, within thirty years the indigenous populations there had been drastically reduced. So drastically, that in Santo Domingo, second largest of the Caribbean Antilles, less than one thousand 'Indians' remained alive by 1519, according to the testimony of Bartolomeo de las Casas, a Spanish monk who came with the conquistadors and was an eyewitness to the conquest.

In his book, *The Colombian Exchange*,[2] the historian Alfred W. Crosby Jr examines the effects of European conquest on the civilisations of both the Old and the New Worlds. He quotes another of the earliest historians of the Americas, Gonzalo Oviedo, who estimated that a million 'Indians' lived on Santo Domingo when the Europeans arrived to settle. "Of all these and of all those born afterwards, there are now believed to be at the present time in this year of 1548, five hundred persons, children and adults, who are natives and are the progeny and lineage of those first."

What Happened to All the Rest?

In the Caribbean and in Mexico, Peru and Colombia, the great Spanish empires of Central and South America, smallpox and other diseases brought over by the Europeans killed the 'Indians' by the millions.

Relatively small Spanish expeditions were able to conquer huge empires because the native populations were swept away by diseases to which they had never been exposed and to which they had no immunity. In North America the advent of the English and the French had the same deadly effect on the native populations. Smallpox and influenza killed Europeans too, but many had partial or total immunity to these diseases, having lived with them from birth in Europe.

Toribio Motolina[3] said that in most provinces of Mexico "more than one half the [native] population died, in others the proportion was a little less; they died in heaps, like bedbugs." More than a hundred years after Motolina, a German missionary wrote in 1699 that the Indians "die so easily that the bare look and smell of a Spaniard causes them to give up the ghost".[4]

The destruction of the American Indian populations and cultures has meant an incalculable loss to human ethnic and cultural diversity. It was the Indians of the Caribbean who gave us words like barbecue, canoe, hammock and hurricane and crops like corn, potatoes and tomatoes. The people of ancient Egypt, the pyramid builders, seem a very long time away; the Maya, Aztecs and Incas, who also built pyramids and played games very much like basketball and jai alai, seem almost like neighbours.

Destroying Africa

The sense of loss is almost palpable too in relation to the lost civilisations of Africa, destroyed by the slave trade. The practical extinction of the

'Indians' meant that massive importations of labour would be necessary to work the plantations and mines of the New World for the enrichment of Europe. The Europeans, judging it impossible for large numbers of Europeans to survive in the inhospitable Indies, turned to Africa for labour. Slave labour was supplied to the American plantations by transplanting whole populations – perhaps 25 million people over three hundred years – from their homes in Africa.

Tearing them from Africa meant the collapse of many ancient cultures which disappeared almost without trace, further impoverishing mankind's cultural diversity and robbing Africa of the populations and the civilisations it required for its own development. Although the Europeans found large quantities of gold in the Americas, the precious metal was – over the centuries – never as important as the sugar, cotton and rubber extracted from the plantations of the New World nor as lucrative as the slave trade and slave labour which made it all possible. The new products from the West soon produced a demand in Europe for new methods to refine the raw materials imported at such cost.

The 'Conquest' of Nature

The Industrial Revolution transformed humanity's relationship with the earth and all other species. In the name of progress and economic prosperity, mankind began to attack the life-sustaining capacities of Earth itself. Resources began to be used faster than they could be replaced naturally.

The steam engine rapidly evolved into the prime mover of mankind, replacing draft animals such as horses, oxen, camels and buffaloes. Steam gave power to locomotives, factories and steamships, tractors and other machines. Chemical fertiliser began to replace animal manure. Eventually, herbicides began to replace hoes and machetes.

The new technologies made possible a new age of rapid economic and social advance. Great prosperity began to exist alongside great poverty and misery. Factories demanded enormous workforces. Mines and plantations developed to satisfy the demands for exotic goods. Men, women and children in Europe, as well as in the Americas, became the lifeblood of industry, the foundation of empires of state and finance. Earth-moving machines, some bigger than houses, made possible the more efficient brutalising of the earth and its varied life forms.

At first, few realised that there was a heavy price to be paid for 'progress' – for confectionery, for disposable diapers, for the consumer society, 'planned obsolescence' and for luxuries which quickly become necessities. Famine, war, sweated labour and pollution brought misery to millions in the wake of the Industrial Revolution. During the seventeenth and eighteenth centuries, Europe was plagued by war, civil unrest and social upheaval.

Modern technology allows us to more efficiently devastate thousands of square miles of forests and liquidate whole ecosystems, to poison entire seas, damage the world's atmosphere and to so accelerate natural processes as to produce catastrophic storms, floods, droughts and epidemics. In the nineteenth and twentieth centuries, industry began to threaten the very basis of life itself even as it created new empires of politics and finance.

Fruits of 'Development'

The US Environmental Protection Agency (USEPA) estimates that in the United States alone there are about 200,000 abandoned oil wells which are either polluting or actively threaten to pollute freshwater aquifers with supersaturated brine, sulphur, heavy metals and other toxins – and radioactivity. It is estimated that it will cost US$4,000 per well to plug them properly. The state of Texas has embarked on a programme of plugging its 50,000 abandoned wells at the rate of 1,000 wells a year. In the state of Oklahoma, the plugging job is expected to take one hundred years, twice as long as in Texas. The petroleum industry is itself not much more than a hundred years old.

It was not until the middle of the twentieth century that humanity as a whole began to appreciate how much damage it was doing to itself and to the rest of the living world. During the Industrial Revolution and after, T.R. Malthus, H.D. Thoreau and others warned that humanity was abusing its heritage. But it was only within the second half of the twentieth century that these warnings began to make news and to be taken seriously by more than a few scientists and conservationists.

Declaring War on Insects

New processes, new alloys, new medicines, new insecticides, new man-made miracles seemed to promise a limitless golden future for mankind.

Information media boasted that progress was infinite.

TIME magazine, in its 12 June 1944, D-Day edition, breathlessly announced a new breakthrough:

> Censorship was lifted last week from one of the great scientific discoveries of World War II. It is an insecticide called DDT. DDT stopped a typhus epidemic in Naples. It promises to wipe out the mosquito and malaria, to liquidate the household fly, cockroach and bedbug, to control some of the most damaging insects that prey on the world's crops.

After cataloguing DDT's promised miracles *TIME* made one disclaimer:

> DDT is not a kill-all. Against two of the most common US crop destroyers, the Mexican bean beetle and the cotton boll weevil, DDT has proved disappointing. Man has not yet won his war with the insects.

'Humanity in the Crossfire'

In 1962, a 50-year-old American marine biologist named Rachel Carson wrote a book called *Silent Spring*[5], in which she exposed and denounced what she called "the current vogue for poisons" to control insect pests.

> These sprays, dusts and aerosols are now applied almost universally to farms, gardens, forests and homes – non-selective chemicals that have the power to kill every insect, the 'good' and the 'bad'. To still the song of birds and the leaping of fish in the streams, to coat the leaves with a deadly film, and to linger on in soil – all this though the intended target may be only a few weeds or insects. Can anyone believe it is possible to lay down such a barrage of poisons on the surface of the earth without making it unfit for all life? They should not be called 'insecticides' but 'biocides'.
>
> The whole process of spraying seems caught up in an endless spiral. Since DDT was released for civilian use, a process of escalation has been going on in which ever more toxic materials must be found. This has happened, because insects, in a triumphant vindication of Darwin's survival of the fittest, have evolved super races immune to the particular insecticide used, hence a deadlier one has always to be developed – and then a deadlier one than that. Thus the chemical war is never won, and all life is caught in its violent crossfire.

Some would-be architects of our future look forward to a time when it will be possible to alter the human germ plasm by design.

But we may easily be doing so by inadvertence, for many chemicals, like radiation, bring about gene mutations. It is ironic to think that man might determine his own future by something so seemingly trivial as the choice of an insect spray . . .

Or by the purchase of a plastic mug, as we shall soon see.

And Now, the Bad News

What Rachel Carson feared is now known to have happened.

Chemical warfare against plants and insects has begun to affect human, animal and plant genetics and will continue to affect them, perhaps for centuries. Albatrosses, living thousands of miles from land, are found to have DDT in their tissues, although the use of DDT was ended worldwide many years ago.

In humans, mothers' milk around the world is laced with chemical residues. Since some of the new organic biocides can cross the placental barrier, new-born children are found with high concentrations of pesticides in their tissues. In the American state of Michigan, in 1973, polychlorinated biphenyls (PCBs and PBBs) used in the manufacture of electrical appliances, were accidentally mixed with animal feed. Although thousands of cattle and chickens were destroyed, contaminated meat caused extended illness in hundreds of farm families and the long-term physical contamination of most residents of the state of Michigan. The breast milk of women who ate the meat contained high levels of PBBs and they produced sons with testicular malformations and undersized penises.

"The control of nature," Rachel Carson said, "is a phrase conceived in arrogance, born of the Neanderthal age of biology and philosophy, when it was supposed that Nature existed for the convenience of man."

Miss Carson's book alarmed many people, including heads of governments. The president of the United States, John Kennedy, set up a panel of experts to report on the dangers of new technologies, and new rules began to be made to control the indiscriminate use of pesticides and other man-made chemicals. And, as Rachel Carson had pointed out, many of these dangerous chemicals were quickly becoming ineffective, as natural organisms began to develop defences against them and were becoming even more deadly than before.

The Bad News Gets Worse

Thor Heyerdahl, the Norwegian anthropologist, voyaging across the Atlantic on his papyrus rafts Ra I and Ra II, found the Sargasso Sea in 1969–70 to be a wilderness of floating plastic garbage which was indiscriminately killing and maiming marine life.

All over the world, fishermen were finding there were fewer fish to catch – the North Atlantic cod fishery, the world's largest, almost disappeared. The cause – overfishing on an industrial scale.

Scientists in the 1980s confirmed that a 'hole' had appeared in the Earth's protective shield of ozone gas in the stratosphere over Antarctica. The hole was caused by industrial chemicals – chlorofluorocarbons or CFCs – reacting with and destroying ozone. Ozone, a gas, acts as a filter high above the Earth, screening ultraviolet solar radiation reaching the surface of the planet and preventing overdoses which damage all living things, plants, animals and people. The consequences of ozone depletion are increased sunburn, skin cancer and cataracts of the eyes, faster ageing of the skin and impairment of the human body's immune system. Excessive ultraviolet radiation can destroy the chlorophyll in plants, inhibiting photosynthesis and therefore accelerating the depletion of oxygen in the atmosphere. Later, another hole was discovered in the Northern Hemisphere, and like the first, the hole kept growing.

The evidence was mounting that human lifestyles were increasingly becoming death-styles; that humanity was drowning in 'progress' and being poisoned by 'development'.

Even more alarming, it soon became unmistakable that not only was the atmosphere becoming dirtier, but the planet itself was becoming warmer because of the so-called greenhouse effect. The Swedish chemist Svante Arrhenius had predicted, in 1896, that increasing industrialisation and the increasing discharge of carbonic acid and carbon dioxide into the air would cause the average temperature of the earth to rise. He predicted that if CO_2 levels were to double, the average surface temperature of the Earth would rise by 4 to 8° Celsius, with catastrophic implications for the planet's climate and its life forms.

The problem with ozone depletion and global warming is that they are both processes with enormous built-in inertia. Even if we stopped producing CFCs and reduced the level of atmospheric pollution of all kinds this minute, the effect of what we do today would not be felt for decades.

The warming, for instance, will continue long after the last man-made fire on earth is cold.

The threat from global warming is manifold. If the Earth generally warmed by 4°C, icecaps at the poles would melt, raising sea levels with disastrous effects on low-lying land masses. But warming would also cause all the ocean's waters to expand in volume, because cold water is denser and takes up less space than warm water. The result would be to raise sea levels even higher than the increase caused by melting ice. In addition, the higher temperatures will shift climatic zones, wiping out certain crops in certain areas, threatening famines and provoking more frequent, freakish and violent extremes of weather – hurricanes, floods, droughts, ice storms, cold spells and heat waves. There is some evidence that this process has already begun. It is believed that global warming helped precipitate the worldwide bleaching of corals in the late 1980s, with disastrous results on beaches and other vulnerable seacoasts. Another effect is in the increasing violence and frequency of Atlantic hurricanes during the past decade, a period which would ordinarily have been expected to be relatively peaceful.

The Bomb and After

The obliteration of Hiroshima and Nagasaki by atomic bombs in 1945 impressed and horrified humanity. The threats and tensions of the Cold War gave rise to a nuclear arms race in which the big powers strove to devise ever more devastating and efficient weapons of mass destruction.

Nuclear weapons testing and the attendant fallout of radioactive debris, contaminating even mothers' milk, gave rise to increasingly vocal anti-nuclear and anti-war movements. And speculation about projects to use atomic explosions to blast new sea level canals between the Atlantic and Pacific Oceans, or to divert Arctic rivers, began to convince many people that the threat of human extinction was real and urgent. Instead of guaranteeing security, the arms race and its technologies were pro-ducing ever more dreadful risks.

Besides which, the global spending on arms was pauperising poorer countries and becoming an increasingly insupportable burden on even the most developed countries. It was, in fact, partly responsible for the collapse of the Soviet Union.

The process of détente was accelerated and simultaneously complicated by the dissolution of the Soviet Union and the creation of several new sovereign nuclear states. Fortunately, only one or two possessed the capacity to wage an actual nuclear war.

During the Cold War, nuclear tensions between the super powers had been reflected in smaller but equally destructive arms races between poorer countries. These contests and conflicts consumed an increasing proportion of their wealth and of the world's non-renewable resources. The legacy includes more than 200 million landmines in Africa and Asia, killing and maiming men, women and children indiscriminately, long after the wars which sowed them had been all but forgotten.

Environmental consciousness was growing all over the world, and especially in North America, in Europe and in the Far East. In India, peasants organised to protect their forests from logging companies by 'hugging' trees marked for cutting. The Chipko Movement, as it became known, began as a protest by a few village women against the logging of forests in their neighbourhoods. The movement was a grassroots one, basically unorganised, but the idea caught the imagination of the peasants and soon spread to involve the imagination of the whole of India. Chipko, according to some writers, helped redefine the very notion of development.[6]

The First Earth Day

In April 1970, thousands of Americans came out to march and to demand an end to environmental pollution. "Cost should not be a consideration," they declared, because there was "a moral imperative to leave the Earth better than we found it." It was the first Earth Day. In the United States, environmentalists and conservationists brought pressure to bear on the government for new laws to regulate development in the national interest. In the US, the National Environmental Policy Act of 1971, the Clean Water Act of 1972 and the creation of the Environmental Protection Agency were all important milestones on the road to sustainable development. In Europe, the so-called Green movement developed political parties, some of which managed to enter parliaments, notably in Germany. In Europe, agitation has led to legislative action to protect the environment.

As Rachel Carson said nearly forty years ago:

The road we have long been travelling is deceptively easy, a smooth superhighway on which we progress with great speed, but at its end lies disaster. The other fork of the road – 'the one less travelled by' – offers our last, our only chance to reach a destination that assures the preservation of our earth.

Even before Rachel Carson wrote her book, there had been ominous warning signs that the golden road to progress was paved with environmental degradation. In December 1952, a killer 'smog' precipitated by particulates from coal-burning fireplaces killed nearly 4,000 people in London, England. The disaster led to the banning of coal-fires in London and helped concentrate attention on environmental health. But frightening environmental disasters began to occur more frequently.

Major Environmental Disasters

In Japan, in the seaside village of Minimata, in the 1960s, children began to be born with dreadful birth defects, crippled, disfigured and brain damaged. The defects were traced to a plastics factory, mercuric effluent from which had been ingested by fish and shellfish eaten by the Minimata villagers. Thousands of people suffered irreversible neurological damage; nearly 1,400 have died.

In Brazil, a similar, perhaps more dangerous environmental disaster is in the making. There, between 1980 and 1995, more than 2,000 tons of mercury have been dumped into the waters of the world's largest river, the Amazon. It is estimated that another 200 tons are annually added by gold miners who use mercury to separate gold from the earth in which it is found.

In Guyana in 1995, the world's largest cyanide spill contaminated rivers when a gold-mining waste dam broke. The consequences of this accident are still unknown.

In the early 1970s in the US state of Virginia, the pesticide kepone contaminated the James River and both the river and parts of Chesapeake Bay were closed to fishing between 1975 and 1980.

A member of the dioxin family, tetrachlorodibenzo-p-dioxin (TCDD), is a contaminant in the widely used herbicide 2,4,5-T, used in agriculture all over the world for nearly forty years. Under its code name, Agent Orange, 2,4,5-T was used by the US military to defoliate large areas of South Vietnam in an attempt to deny vegetative cover to the nationalist Viet Cong guerrillas.

Although no precise figures are available, it is widely believed that the herbicide caused innumerable animal deaths and is believed to have caused cancers and birth defects among Vietnamese exposed to it. American veterans of the war have also complained of disabling illnesses and many of their children have been afflicted by serious birth defects and cancers which have been attributed to Agent Orange. Dioxins have also entered the environment through the use of tainted oil as a roadway dust suppressant. The town of Times Beach, Missouri, was abandoned in 1983 because tests indicated that highway spraying had left high levels of dioxin in the soil.

Toxic Soil, Burning Rivers

In New York state, land used by a chemical company for twenty years as a toxic waste dump was filled and given in 1953 to the city of Niagara Falls. A new school and many houses were built on it. In 1971, toxic liquids began leaking through the clay cap that sealed the dump, and the area was contaminated by at least eighty-two chemicals, including a number of carcinogens: benzene, some chlorinated hydrocarbons, and dioxin. Extremely high birth-defect and miscarriage rates developed, as well as liver cancer and a high incidence of seizure-inducing nervous disease among the neighbourhood children.

On 3 December 1984, Bhopal, a city of more than a million people and capital of the Indian state of Madhya Pradesh, was the scene of the worst industrial accident in history up to that time. Some 2,500 persons died and as many as 200,000 were injured when toxic methyl isocyanate gas leaked from a pesticide plant there. The tragedy raised serious questions about the safety of the chemical industry worldwide, and the legal and moral responsibilities of multinational corporations operating in Third World countries.

Nuclear accidents also caused severe damage to human beings, other life forms and to property. The world's first known serious nuclear accident occurred in 1957 at a British weapons production plant, Windscale 1, an air-cooled graphite reactor. During start-up operations the plant caught fire and volatile fission products, most notably iodine and cesium, were released into the environment around the plant. The Windscale plant and its twin unit were eventually shut down.

The most notorious nuclear accidents were those at Three Mile Island in the United States in 1973 and Chernobyl in the Soviet Union in 1986.

A small leak in the Three Mile Island nuclear reactor, near Harrisburg, Pennsylvania, led to a partial meltdown of the nuclear core. The molten fuel and cladding fell into the bottom of the vessel which, fortunately, was strong enough to contain the debris, preventing a more dangerous situation. Although this accident provoked worldwide apprehension of apocalyptic nuclear disaster, it was insignificant compared to the accident at Chernobyl, in the Ukraine, thirteen years later.

At Chernobyl, human error caused an explosion which killed thirty-one people immediately or shortly thereafter and directly injured more than five hundred others. Fallout from the Chernobyl incident was reportedly twenty times greater than the combined effect of the atomic bombs dropped on Hiroshima and Nagasaki by the US in 1945.

The force of the explosion and fire carried much of the radioactivity away from the site to relatively high altitudes, where it spread across the Northern Hemisphere. Fallout contaminated the western Soviet Union and portions of Europe, where steps were taken by several nations to try to protect food supplies.

The authorities have admitted that several million people in the Ukraine and Belorussia are still living on contaminated ground. The incidences of thyroid cancer, leukaemia and other radiation-related illnesses are higher than normal among this population. In addition, it is believed that radiation-induced mutations in human, animal and plant genes may yet produce monstrous and unpredictable results in the offspring of those affected.

But even more alarming discoveries began to be made:

> Beginning in the 1950s, bizarre and puzzling abnormalities began to surface in animal populations in different parts of the world – in Florida, the Great Lakes, in California, in England, Denmark, the Mediterranean, and elsewhere. Many of the disturbing wildlife reports involved defective sexual organs and behavioral abnormalities, impaired fertility, the loss of young, or the sudden disappearance of entire animal populations. In time, the alarming reproductive problems first seen in wildlife touched humans, too.
>
> Each incident was a clear sign that something was wrong, but for years no one recognised that these disparate phenomena were connected. While most incidents seemed linked somehow to chemical pollution, no one saw the common thread.[7]

It now seems more than probable that hundreds of the most familiar chemicals of the postwar age – PCPs – used in the manufacture of electronics, pesticides such as endosulfan and atrazine, polycarbonate plastic found in many baby bottles, water jugs and soft drink bottles, and chlorine compounds that bleach paper – resemble the human sex hormone oestrogen. Plastics once thought to be safely inert now appear to carry a dangerous freight of active hormone-disrupting chemicals. Their molecular structures are so similar to oestrogen that they fit into the same receptors in human tissue. These chemical mimics can trick the body into turning on or modifying certain biochemical pathways, especially those in the reproductive system, with disastrous results for human and animal genetics.

The result is a worldwide reduction in male sperm counts, distortion of sexual development in both males and females and breast cancers and other serious diseases.

Petroleum, the lifeblood of modern development, has been the cause of several environmental disasters. In 1979, a major disaster was caused by the grounding of the tanker *Amoco Cadiz* in the English Channel. Ten years later, the tanker *Exxon Valdez* released 41 million litres (9.11 million gallons) of oil when it ran aground in Alaska's Prince William Sound. Thousands of seabirds were killed and enormous damage done to the wildlife of the until then relatively unpolluted sub-Arctic Alaskan waters.

Limits to Growth?

The Club of Rome is an informal international organisation, set up in 1968 to promote greater understanding of the interdependence between global economic, political, social and natural systems. The Club of Rome sponsored a research programme at the Massachusetts Institute of Technology on the prospects for world development. The study, published in 1972 as *Limits to Growth*, was an environmental bombshell in the world of international political economy, declaring that world economic growth could not be sustained much beyond the end of this century, particularly because of the depletion of non-renewable resources and the effect of population pressure.

The so-called Green Revolution, concentrating on producing uniform, high-yielding varieties of crops, ultimately failed to produce the marvellous benefits predicted. The Green Revolution's capital-intensive, expansive character increased the inequality of wealth distribution, reduced biodiversity by concentrating on a few, uniform crops, increased the dependence on chemical pesticides and fertilisers and cleared enormous areas for short-lived farms. In Mexico, one researcher concluded that it worsened the absolute living standards of Mexico's poor rural population, reduced domestic production of food crops and raised the price of food, resulting in increased hunger and greater concentration of poverty in the countryside.[8]

As the Cold War began to evaporate, a growing torrent of books, magazine and newspaper articles, television news stories and documentaries began to alert people to the dangers of exploitative economic growth. Important areas of national and international economies depended on the arms race and the end to the superpower arms race emphasised the need for new strategies for development.

Our Common Future?

The Stockholm Conference on the Human Environment in 1972 began to lay the foundations for an explosion of environmental consciousness around the world and the coordination of work toward international agreement on how humanity might remedy its past mistakes.

A decade later, in 1983, concerned by the mounting evidence of increasingly rapid deterioration of the biosphere, the United Nations set up the World Commission on Environment and Development to report on the problem and suggest possible solutions. This study group became known as the Brundtland Commission after its leader, Mrs Gro Harlem Bruntdland then Foreign Minister, later Prime Minister of Norway.

The Commission's report, *Our Common Future*, as it was called, was published in 1987. It was a major landmark on the road to sustainable development, warning that humanity would have to change its ways or else face unacceptable levels of environmental degradation leading to widespread human suffering. The Brundtland Commission said that while the global economy had to meet the needs of the human population for improved standards of living, it had to fit within ecological limits or else economic development could not be sustained and its benefits would not last. The United Nations, after considering the Brundtland Report, began to prepare a world conference to take decisions on how the world's people could plan for survival into the twenty-first century and beyond.

In 1992, the leaders of most of humanity met in Rio de Janeiro, Brazil, to agree for the first time on a worldwide effort to devise strategies for sustainable development. This summit meeting, the United Nations Conference on Environment and Development, was also called the Earth Summit. It was the first time that the leaders of most of the world's people were meeting to recognise – formally and officially – that the world's people could choose between surviving cooperatively or perishing independently.

The UN Conference on the Environment and Development was also the first occasion in human history on which it could be said that humanity as a whole reached a formal consensus on anything.

There had been major conferences before, to end wars, to allocate the spoils of war, cutting up continents and dividing peoples. This was the first time that most of the world's people were represented at a conference in which the world was deciding on its future.

Attitude and Lifestyle Changes

It had taken a very long time for the world to recognise that conventional 'development' could not be commandeered for the benefit of only a few of the world's people, leaving the rest poor, miserable and hopeless. In conventional development, the world was itself becoming a poorer place to live, degraded environmentally and increasingly hostile to human existence and human happiness. The challenge was to find ways in which development could be made to benefit all human beings while preserving basic resources for our children and their descendants. Poverty should be eliminated and the vast disparities in incomes should be reduced, because poverty – while an important contributor to environmental degradation – is also the result of environmental degradation.

Development cannot be stopped. It must be made to change course, to become productive and sustaining instead of exploitative and destructive.

Sustainable development means not just a change in industrial attitudes, but also, and more important and certainly more difficult, changes in the lifestyles of most people – the rich as well as the poor. Above all, as Jamaican professor Elizabeth Thomas-Hope has said: "Green development is not about the way the environment is managed, but about who has the power to decide how it is to be managed."[9]

Agenda 21 is the roadmap drawn up by the UN to guide government, business and ordinary people into the next century. But Agenda 21 is only one of five major documents produced by the Earth Summit. These documents together provide a framework for individual, community, corporate and government action which will change the way we live. Hopefully, it will help to guarantee those who come after us a more hospitable world environment – where we live, how we live, how we work and how we organise our lives.

The five documents of Rio carry the message that we human beings have no future unless we have a future in common. We either stand together or fall separately. Agenda 21 is the master guideline for action towards sustainable development. The other documents of Rio are:
• The Rio Declaration on Environment and Development
• Statement of Principles on Forests
• The Convention on Climate Change
• The Convention on Biological Diversity

All these documents are summarised as appendices to this book. Also appended is a summary of the decisions taken at the Conference of Small Island Developing States (SIDS) in Barbados in 1994.

Notes

1 Pain, Stephanie. 1994. "Rigid cultures caught out by climate change", *New Scientist* 5 March, p. 13.
2 Crosby, Alfred W. Jr. 1972. *The Colombian Exchange*. Westport, Conn.: Greenwood Press.
3 Motolina, Toribio, quoted by Crosby, *The Colombian Exchange*.
4 Oviedo, quoted by Crosby, *The Colombian Exchange*.
5 Carson, Rachel. 1962. *Silent Spring*. New York: Houghton Mifflin.
6 Guya, Ramachandara.1988. "Ideological trends in Indian environmentalism", *Economic and Political Weekly* 23 (49).
7 Colborn, Thea, Dianne Dumanoski, and J.P. Myers. 1996. *Our Stolen Future*. New York: Dutton.
8 Hewitt de Alcantara, Cynthia. 1976. *Socialeconomic Implications of Technological Change*. New York: UN Research Institute for Economic Development (quoted in Bruce Rich. 1994. *Mortgaging the Earth*. Boston: Beacon Press).
9 Thomas-Hope, Elizabeth M. 1996. *The Environmental Dilemma in Caribbean Context*. Kingston: Grace Kennedy Foundation Lecture.

The Declaration of Rio

Principles of Sustainable Development

The nations meeting at Rio de Janeiro for the Earth Summit adopted a set of principles – Agenda 21 – to guide future development. These principles define the rights of people to development, and also define their responsibilities to protect the world environment which is the common heritage of mankind.

Agenda 21 builds on the principles adopted in 1972 in the Stockholm Declaration of the UN Conference on the Human Environment.

Long-term economic progress – sustainable development – is practicable only if it is linked with environmental protection and is possible only if the nations of the world establish new and equitable global partnerships between governments, peoples and the key sectors of their societies.

International agreements are required to protect the integrity of the global environment and of the process of sustainable development.

The basic principles are:
- The right of every human being to a healthy and productive life in harmony with nature.
- Development must not endanger the prospects either of those now living or those not yet born.

- While sovereign states have the right to exploit their own resources, they should do so cautiously, with full regard for the interests of their own people and of people outside their countries.
- Nations should develop international laws to provide compensation for damage caused outside their states by activities inside their states.
- Nations should adopt a precautionary approach to protect the environment from the effects of development. Effective measures must be adopted wherever there is the threat of irreversible damage, despite any lack of scientific certainty that such damage will occur.
- Environmental protection must be an integral part of the development process and cannot be considered in isolation from it.
- The eradication of poverty everywhere and the reduction in disparities in living standards are essential.
- Nations must cooperate to protect and restore the health and integrity of the global ecosystem. Developed countries acknowledge the responsibility they bear in the international pursuit of sustainable development, in view of the pressures their societies place on the global environment and because of the technological and financial resources at their command.
- Nations should eliminate unsustainable patterns of production and consumption and promote appropriate population policies.
- Nations shall facilitate the full participation of their citizens in environmental decision making.
- Nations shall cooperate to enact effective environmental protection laws governing liability for victims of environmental damage.
- Nations should cooperate in developing an open international economic system to promote worldwide economic well-being.
- The polluter should pay for his pollution.
- Nations should share scientific knowledge for sustainable development.
- The fullest participation in the decision making and development processes is essential to sustainable development.
- Since warfare is inherently destructive of sustainable development, nations are enjoined to respect international laws protecting the environment in times of war.
- Peace, development and environmental protection cannot exist in isolation from each other. They are interdependent and indivisible.

Appendix Two

Agenda 21

A Roadmap for Sustainable Development

Section I

Social and Economic Dimensions

International Cooperation; Combating Poverty; Changing Consumption Patterns; Population and Resources; Health and Environment; Housing; National Decision Making.

Section II

Resource Conservation, Management

Protection of the Atmosphere; Land Use Management; Deforestation; Drought; Mountains Ecosystems; Rural Development: Biological Diversity; Biotechnology; Sea & Ocean; Freshwater; Toxic Chemicals; Hazardous Wastes; Solid Wastes, Sewage; Radioactive Wastes.

Section III

Strengthening the Role of Major Groups

Women & Development; Children & Youth; Indigenous Peoples; Non-Governmental Organisations; Local Government; Workers & Trade Unions; Business & Industry; Scientists & Technologists; Role of Farmers.

Section IV

Means of Implementation

Financing Sustainable Development; Transfer of Technology, Cooperation, Capacity Building; Scientific Research & Evaluation; Education, Public Awareness & Training; Building Capacity for Sustainable Development; International Institutional Arrangements; Legal Instruments; Information for Decision Making.

AGENDA 21
Section 1
Social and Economic Dimensions

1. Environmental protection must be an integral part of development and development should be aimed at eradicating poverty, and reducing extreme disparities in living standards.

   Nations should share knowledge and innovative technologies in order to help each other attain sustainable modes of development.

   Democratic participation: In all of this every human being should have the right to participate democratically, having equal rights in the decisions and development which may affect them.

   Polluters should pay for their pollution.

   Governments, of course, have the major role in achieving sustainable development because they must develop national strategies, plans and policies which will guide sustainable development and allow full participation in that development.

   Poorer countries must be helped. It is recognised that if poorer countries are to achieve sustainable development they must get help from the richer countries, in finance, technology, cooperation and information.

International Cooperation

2. International cooperation is an essential component of Agenda 21 which envisions a partnership between the world's nations in the construction of an efficient and equitable global economy, in which there eventually will be no haves and have-nots.

   The international community must cooperate to make the terms of trade between nations more just, eliminating artificial restrictions, subsidies and other devices which restrict trade and prevent poorer countries from realising their full potential.

Combating Poverty

3. Combating poverty is a major priority of Agenda 21. If poverty is to be eradicated, nations must bring all their people into the development process, paying special attention to women and young people who are everywhere marginalised. Poverty has so many causes that there can be no single simple solution.

    Eradicating poverty is important in preserving our natural resources. The world can no longer tolerate widespread unemployment because it not only wastes people but promotes the kind of desperation which helps to destroy valuable natural resources.

    Humanity must change its behaviour if the world is to eradicate poverty and reduce the disparities in wealth between people.

Changing Consumption Patterns

4. Changing consumption patterns is a key ingredient in managing developmental problems. Human beings need to reconsider their standards and concepts of wealth and prosperity. We need to understand that we can only survive if we understand that the earth does not have unlimited carrying capacity. The human race, in other words, must learn to live within its means.

    Dangerous and wasteful patterns of consumption must be abandoned if we are to satisfy the basic needs of the poor.

    Western lifestyles depend on production and consumption patterns which are not sustainable. These lifestyles are, however, the envy of much of the rest of the world and if we are to avoid disaster, all of us must recognise that we need to produce the more essential goods, reducing waste and consumption for show.

    Pricing and taxation policies which penalise wasteful production of goods and wasteful consumption of natural resources must be introduced and adopted by all countries.

Population and Resources

5. Population policies are integral to sustainable development. Countries need to know how many people their territory can support and to plan and work within those limits to provide the jobs, good health and quality of life to which all people are entitled. The carry-

ing capacity is the ability of the resource base to support and provide for the needs of the people without becoming depleted.

The status of women must be raised if countries are going to be able to work within the limits of the resource base. Countries need to provide women with reproductive health programmes and the information and means to plan family size. If the family is made secure, there will be a reduction in the urge to guarantee security by producing more children.

Health and the Environment

6. Human health depends on healthy environments, including clean water, sanitary waste disposal and a healthy and adequate diet. The world needs to campaign against ill health and the conditions which promote unhealthy living. Health care should be adapted to local needs.

   National public health programmes should include a National Health Watch to deal with communicable diseases. Community based health care systems should be promoted to give people the interest and the knowledge to deal with their own basic health needs. If possible, traditional knowledge and experience should be integrated into national health systems.

   The overall strategy is to achieve health for all by 2000 and health strategies should be designed to give priority to those in greatest need.

   Action must be taken to minimise air pollution, reduce the use of dangerous pesticides and ensure the sanitary disposal of solid waste.

Adequate Housing

7. Nations should aim to provide adequate shelter for all, to improve the management of human settlement, to promote sustainable land use management and the integrated provision of infrastructure – water, sanitation, drainage and solid waste management.

   By the year 2000, it is estimated that 3,000 million people – half the world's population – will be living in cities.

   To make urban life more sustainable, governments should make it easier for the poorest people to get access to land, credit and low cost building materials.

Building programmes should emphasise local materials, energy efficient designs and labour-intensive technologies to employ more people. Nations should pay special attention to providing sustainable energy resources for urban households and adequate transportation.

A 'culture of safety' should be developed to protect those vulnerable to disasters and the construction sector should be encouraged to turn to sustainable building techniques. Rural people should be discouraged from moving into the cities by improving rural living conditions and encouraging medium-sized cities that create employment and housing.

National Decision Making

8. National decision making should be restructured to integrate economic, social and environmental factors in the planning and execution of development.

AGENDA 21
Section II
Conservation & Management of Resources

Protection of the Atmosphere

9. We need to understand the processes at work in the atmosphere and to be able to predict possible and probable changes and their effect on human life.

   The international community needs to cooperate in research, the development of early detection systems concerning changes in the atmosphere, to identify and control atmospheric pollutants.

   The basic objective is to identify and reduce adverse effects on the atmosphere from the energy sector and to promote the sustainable and healthy use of energy of all kinds.

   Nations are urged to comply with international agreements, especially those relating to the discharge of gases which damage the ozone layer.

   Nations should promote sustainable uses of energy and programmes to minimise adverse impacts on the atmosphere from all sources.

Land Use Management

10. Since land use management is an important component of sustainable development, there should be an integrated approach to facilitate the allocation of land uses to those projects which produce the greatest sustainable benefits.

Combating Deforestation

11. Deforestation is one example of bad land use management which produces adverse effects on the whole environment. Better management must be introduced to protect forests and enhance their functions as economic resources, as well as 'sinks' for greenhouse gases, reservoirs of biodiversity and all the other ecological, economic, social and cultural functions of forests.

Combating Drought and Desertification

12. Fragile ecosystems urgently need protection to preserve their unique features and resources.

    Fragile ecosystems include deserts, semi-arid lands, mountains, wetlands, small islands and coastal ecosystems.

    Desertification affects one sixth of the world's population and one-quarter of the world's land area. Programmes must include better management of these systems and development of economic activities which will help protect these systems while improving the standard of living of those who depend on them.

Mountain Ecosystems

13. Mountains are the areas most sensitive to climatic change and are extremely vulnerable to human interference. Mountains and hill-sides, because of their climatic gradients, contain many ecosystems and need to be managed to preserve these systems particularly by protecting watersheds, controlling soil erosion and preserving their unique characters.

Sustainable Rural Development

14. Rural development and agriculture require major adjustments, particularly because increasing populations are putting increasing stress on the capacity of farmers to grow more food. National policies should aim for food security through the better management of agricultural production and the harvesting, storage and processing of agricultural products to minimise waste.

    In this area particularly, there is urgent need to involve the local inhabitants in the management of agricultural and rural resources with better extension services, increased access to capital and better security of tenure. Rural development should be enhanced by diversification of production and systems and by non-farm employment, infrastructure development, planning information and education.

    Plant genetic resources must be conserved to provide resources for future needs. Food plant diversity must be protected. Comprehensive programmes must be undertaken to improve food production.

Protecting Biological Diversity

15. Biological diversity – the variability and variety of genes, species, populations and ecosystems – must be maintained to protect the health of the planet. Programmes are recommended to stop habitat destruction, over harvesting, pollution and the introduction of foreign plants and animals which sometimes wipe out important indigenous species.

    National plans must take into account the conservation and sustainable use of biological diversity for the widest sharing of the benefits from the use of biological and genetic resources. All governments should accede to the convention on biological diversity.

    Nations should encourage the sustainable use of biodiversity, and recognise the importance of traditional knowledge in this area. Genetic material in plants and animals has the potential to offer enormous benefits to humanity. The marine environment, increasingly threatened, is particularly rich in species diversity.

Managing Biotechnology

16. Biotechnology must be managed in the interest of humanity. The use of genetic material in plants, animals and microbial systems will continue to lead to many useful products with a variety of uses and benefits. Biotechnology can offer great advances in agriculture and food production, in livestock management, pest control and other areas.

    Biotechnology has the potential for big improvements in human health, and in enhancing the protection of the environment by producing new technologies to repair degraded ecosystems, deal with waste and promote recycling.

    The application of biotechnology must be environmentally sound and not exploitative. It should encourage the development of industries in developing countries based on their genetic resources.

The Seas Around Us

17. Oceans and all water resources require special protection because of their vulnerability to abuse and the variety of living resources they contain. Coastal states should establish high level integrated

management for sustainable development of coastal areas and their marine resources.

Decision making should always include the results of environmental impact assessments. Coastal settlements and the use of the coast by industrial, fishing, recreational and other interests must be regulated.

The coastal environment may be degraded by improper land use, so proper land use behind the coast must be part of the coastal zone management process. Seas and other waters should not be used as dumps and should be protected from degradation by shipping, mining and other offshore activities.

Some particularly sensitive areas may need special protection, particularly because the marine environment is peculiarly vulnerable to climate and atmospheric changes. Much more information and research is required into all areas of the marine and coastal environments.

Already, 60 percent of the world's population live in coastal areas and 65 percent of cities with more than 2.5 million people are on the world's coasts, many of them at or below sea level.

Small island states, most of which are developing countries, must be a special concern both for environment and development. They are rich in biological diversity and have all the problems of coastal states in a much smaller land area. Their resources are limited and they are very vulnerable to global warming, since so many of the people live in low-lying coastal zones. They need international assistance to develop strategies for their peculiar developmental problems and to promote sustainable development within the context of their size and lack of resources.

Freshwater Resources

18. Freshwater resources require urgent protection because they are threatened by watershed destruction, overuse and pollution. Water resource development is a vital component of economic productivity and social well-being.

    Freshwater supplies must be protected, enhanced if possible and managed in the interest of all. Sources of supply must be protected from abuse, overuse and pollution. Flood and drought management must be improved and people must be educated to use water more rationally. Nations must not only protect their aquatic resources but also the living resources within them.

Climate change threatens water resources because higher temperatures will mean not only less rain but also increase the demand for water. Any rise in the sea level will mean saltwater intrusion into estuaries, small island and coastal aquifers, putting low-lying countries at great risk.

Safer Use of Toxic Chemicals

19. Toxic chemicals must be managed more rigorously and the trade in them controlled. The illegal traffic in toxic chemicals requires coordinated international action to determine the safety of all chemicals, to assess the degree of risk posed by these chemicals and to produce guidelines for thresholds of acceptable risk.

Managing Hazardous Wastes

20. Hazardous wastes must be effectively controlled at the point of generation, storage, treatment, recycling and reuse, transport and disposal. The overall objectives include minimising the production of toxic waste, ensuring its safe storage and transport and prohibiting its movement across national boundaries.

Solid Wastes and Sewage

21. Solid wastes and sewage must be managed more effectively. By 2000, all industrialised countries should be operating programmes to stabilise or reduce the amount of waste and all countries should be reducing agricultural wastes. Non-governmental organisations and consumer groups should be encouraged to participate in strengthening national capacities for environmentally sound and resource-producing ways of managing solid wastes.

Radioactive wastes

22. There is a need to ensure that radioactive wastes are safely managed, transported and stored, particularly in countries which are preparing

to go into large-scale use of nuclear power.

Radioactive wastes are generated by nuclear reactors producing electrical energy and by medical, industrial and research operations. About 200,000 metric tons of nuclear waste are generated annually worldwide, some of it low-level waste, but most is highly radioactive and toxic waste.

AGENDA 21
Section III
Strengthening the role of major groups

23. Preamble: none of the objectives, policies and mechanisms agreed to by governments in Agenda 21 are likely to be achieved without the fullest participation of their whole societies. One of the fundamental prerequisites for the achievement of sustainable development is broad public participation in decision making.

    If sustainable development is to be achieved, new forms of public participation are needed.

    The people must take part in environment impact assessment procedures and be informed about and take part in decisions, particularly those which specifically affect the communities in which they live and work.

    Members of the public have the right of access to information held by national governments, including information on products and processes which may have an impact on their lives.

    The UN system must give equal access and representation to all major groups – women, youth, indigenous people, non-governmental organisations, workers and trade unions, business and industry, scientists, technologists and farmers – in any discussions involving Agenda 21.

Women And Sustainable Development

24. The status of women: an important component of Agenda 21 is to give effect to all the various plans and conventions directed towards the full, equal and beneficial integration of women in all development processes. The main aims are to implement the Nairobi Forward Looking Strategies for the Advancement of Women; to increase the proportion of women decision makers, planners, technical advisers, managers and extension workers in the environment and development fields.

    Other major aims include: the formulation by 2000 of a strategy of changes necessary to fully integrate women into all aspects of public life and authority, to ensure that mechanisms are established to guarantee national, regional and international compliance with these aims; to educate both men and women in gender-relevant knowledge and valuation of women's roles; to promote and implement clear national policies and guidelines for the achievement of equality in all aspects of society, including the promotion of women's literacy, education and health.

A Role for Children and Youth

25. Children and youth are critical to the success of Agenda 21. Nearly one third of the world's population are under 15 years old and more than half of all the people in the world are under 25 and four out of five of them live in developing countries.

    Governments must promote dialogue between themselves and youth and take advantage of the fact that youth has unique perspectives which should be taken into account. Every country by 2000 should have at least half of its young people, male and female, either enrolled in or with access to appropriate secondary or vocational education. Nations should make it a priority to reduce youth unemployment and to combat human rights abuses of children and young people, particularly of young women and girls.

    Young people should be a part of every nation's delegation to all UN processes so that they can influence these processes directly.

    Governments should ensure that young people have all the legal protection, skills, opportunities and the support necessary for them to fulfil their personal, economic and social aspirations and potentials.

    All this must be realised in the context that our children will shortly inherit the duty of taking care of the earth.

The Role of Indigenous Peoples

26. Indigenous peoples and their communities must be fully empowered to participate in all the processes of sustainable development. They must be protected in their peaceful enjoyment of their ancestral lands. And it must be recognised by governments that traditional and direct dependence on renewable resources and ecosystems, including sustainable harvesting, continue to be essential to the cul-

tural, economic and social well-being of indigenous peoples. They must be encouraged and assisted in capacity building, so that they may take an increasingly direct control of their lives and participate fully in sustainable development.

The Role of Non-governmental Organisations

27. Non-governmental organisations must be strengthened and encouraged to play a more direct role in the shaping and implementation of participatory democracy. Their independence from government interference should be guaranteed.

The world needs to develop a sense of common purpose as it moves away from unsustainable practices towards development that is environmentally safe, sound and sustainable. The chances of forging such a sense of purpose will depend on the willingness of all sectors to participate in genuine social partnership and dialogue, while recognising the independent roles, responsibilities and capacities of each. The UN system should promote the participation of NGOs in all activities relating to Agenda 21. NGOs should foster cooperation and communication among themselves to reinforce their effectiveness.

Local Authorities

28. Local authorities play a vital role in sustainable development, being the level of governance closest to the people.

By 1996, most local authorities in each country should have undertaken a consultative process with their populations and achieved a consensus on a 'Local Agenda 21' for their communities.

By 1994, representatives of local authorities and cities should have increased their level of cooperation and coordination with the goal of enhancing the exchange of information and experience between local authorities.

Workers and Trade Unions

29. Workers and trade unions are at the forefront of the implementation of Agenda 21, involving as it does transformations at the national and enterprise levels of development. Trade unions are already vital

actors in improving the workplace environment and in the protection of the surrounding environment. The existing network of collaboration among and between trade unions provides important channels through which the message of sustainable development may be carried and implemented. The established basis of tri-partisan collaboration between workers, government and employers, provides an important platform for the implementation of sustainable development. Trade unions and employers should establish the basis for joint environmental policy. Workers should participate in environmental audits at the workplace and play an active role within the United Nations in the implementation of sustainable development.

Business and Industry

30. Business and industry, including transnational corporations, are crucial in the attainment of sustainable development. They require a stable policy regime to operate responsibly and efficiently, and they are major contributors to increased prosperity, increased employment and technological advance. They should be full participants in the implementation of Agenda 21 at the national and international levels, including the UN system.

     Business and industry should recognise environmental management as among the highest corporate priorities and as a key determinant in sustainable development.

     With the cooperation of government, business and industry should aim to increase the efficiency of resource use, including the reuse and recycling of residues and reducing the amount of waste discharge per unit of output, promoting cleaner production.

     Responsible entrepreneurship can play a major role in improving the efficiency of resource use, reducing risks and hazards, minimising wastes and safeguarding environmental quality.

     Enterprises should embrace the concept of stewardship in their management and use of resources.

Scientists and Technologists

31. Scientific and technological communities nationally and internationally must be enabled to make a more open and effective contribution to the developmental process, functioning not simply as expert advisers, but as full participants in the decision making process, bringing their experience as well as their expertise to bear. Partnership between the experts and the public depends on improved communication and better training. The public should be assisted in getting its ideas across to the scientists about how science and technology may be better managed to improve the lives of all.

Important Role for Farmers

32. Farmers, including all rural people and everyone who lives by farming, fishing and forestry activities, comprise the largest single group of workers. Rural activities take place close to nature, adding value to it and producing renewable resources and at the same time are vulnerable to overexploitation and improper management.

Local and village organisations must be strengthened to produce more decentralised decision making; women and other vulnerable groups must have greater and more secure access to the use and tenure of land.

Sustainable farming and related technologies must be promoted, along with policies to encourage self-sufficiency in low-input and low-energy technologies; policies must be developed to provide incentives and motivation for sustainable farming practices; and all farmers and rural workers, male and female, must be encouraged and assisted to help in the design and implementation of policies towards these ends.

AGENDA 21
Section IV
Means of Implementation

Financing Sustainable Development

33. Financial resources and mechanisms: Agenda 21 can only work if there is serious national and international commitment to make it work. International cooperation for sustainable development needs to be strengthened in order to support and complement the efforts of developing countries, particularly the least developed countries.

The implementation of the huge programmes of Agenda 21 will require substantial new and additional financial resources according to sound and equitable criteria and standards. The initial phase will be accelerated by the substantial early commitments of concessionary financing. In general, funding for sustainable development will come from each nation's own public and private sectors.

For developing countries, particularly the least developed, Overseas Development Assistance (ODA) will be the main means of financing.

Developed countries have reaffirmed their commitments to reach the accepted UN target of 0.7 percent of GNP for ODA and several have committed themselves to reaching that target by 2000. The principle of equitable burden sharing among developed countries is recognised.

Funding should be provided with the aim of maximising new and additional resources and all available funding sources and mechanisms should be employed.

Among these resources and mechanisms are: the multilateral banks and agencies and the specialised agencies of the UN; multilateral agencies for capacity building and technical cooperation; bilateral assistance programmes; debt relief; private funding; investment and innovative financing generating new public and private financial resources including debt relief and the transfer of funds from military budgets.

Transfer of Technology, Cooperation, Capacity Building

34. Environmentally sound technologies are 'process and product' technologies and systems which generate little or no waste. Access to these technologies should be easy for all, especially for developing countries.

New measures should promote cooperation and transfer of needed know-how, as well as building up economic, managerial and technological capacity for the effective use of transferred technology.

These transfers should be on favourable, possibly concessionary terms, taking into account the need to protect intellectual property rights, as well as the special needs of developing countries for the implementation of Agenda 21.

Environmentally sound indigenous technology which may have been ignored or neglected should be promoted.

There should be increased emphasis on human resource development, institution building, practical assessments of technology needs and the promotion of long-term technological partnerships between holders of environmentally sound technologies and potential users.

These programmes will require, among other things, the development of international information networks to link national, sub-regional, regional and international systems; establishment of a collaborative network of research centres and collaborative arrangements and partnerships.

Science for Sustainable Development

35. Sustainable development requires taking long-term perspectives, integrating local and regional effects of global change into the development process and using the best scientific and traditional knowledge available.

The development process needs to be continually evaluated to ensure that resource utilisation does mean a reduction of impacts on the Earth's systems. Even so, there will be occasional surprises. This means that good environmental management must be scientifically robust, seeking to open a range of options to ensure flexibility of response to unexpected events. The precautionary approach is essential and there must be better communication between scientists, the policy makers and the public at large.

Nations need to identify the state of their scientific knowledge, especially in areas relevant to environment and development. There needs to be more interaction between scientists and decision makers, using the precautionary approach, to change existing patterns of production and consumption and to gain time for reducing uncertainty in the selection of policy options.

There is a need to generate and apply knowledge to the capacities of different environments and cultures and to use indigenous and local knowledge to achieve sustainable development.

There must be more cooperation between scientists in the promotion of interdisciplinary research programmes and activities, and people should have more say in setting priorities and in making decisions about sustainable development.

Education, Public Awareness and Training

36. Education, public awareness and training are vital to the achievement of sustainable development, particularly in the area of satisfying basic needs – a major priority of sustainable development and Agenda 21.

Education is a process by which human beings and society can reach their full potential. Education is critical for promoting sustainable development and improving the capacity of people to assess and deal with environmental and developmental issues. Education is critical for the achievement of environmental and ethical awareness, and to instil and promote values, attitudes and to distribute skills consistent with sustainable development and effective participation in decision making.

The first priority is to achieve primary education for at least 80% of all children, boys and girls through formal or non-formal education and to reduce the adult illiteracy rate to half the 1990 level. Other objectives are: to develop environmental awareness and to link environmental and developmental education to social education, from primary level to adulthood, among all groups. It is essential to promote the integration of developmental and environmental concepts throughout the educational system and to localise analyses of the causes of global environmental problems. Special attention should be paid to the training of decision makers at all levels.

Countries should set up national advisory bodies on environmental education and the school should become a centre for community education and action on the environment and development. All schools should be encouraged to develop environmental action programmes. It is necessary to maximise information exchange to strengthen environmental education and public awareness.

Nations should support university and other tertiary activities and networks for environmental and developmental education.

Countries should strengthen or establish regional centres of excellence in interdisciplinary research, promoting cooperation and information sharing and dissemination and non-formal educational initiatives by community groups and individuals. Adult education programmes should be promoted and there should be special efforts to foster the education of women in non-traditional activities.

Increasing public awareness: because most people are still unaware of the interrelated nature of all human activities and the environment, there is a need to increase public sensitivity to and awareness of environmental and developmental problems. The aim is to promote broad public awareness of these concerns as part of a global education effort to strengthen attitudes, values and actions compatible with sustainable development.

It is important to emphasise the principle of devolving authority, accountability and resources to the most appropriate level, with preference given to local responsibility and control over awareness building activities.

Countries should establish or strengthen existing bodies for public environment and development information; should encourage public participation in discussion of environmental policies and assessments; should provide public environmental and development information services for raising the awareness of all groups, the private sector and particularly decision makers.

Educational institutions, especially in the tertiary sector, should be stimulated to contribute more to awareness building. The media, popular theatre groups and the entertainment and advertising industries should be stimulated to mobilise their experience in shaping public behaviour and consumption patterns. Such cooperation would increase active public participation in the debate on the environment.

Countries should, in cooperation with the scientific community, employ modern communications methods to maximise outreach, expand the use of audiovisual methods and produce radio and television programmes integrating advanced technologies with folk media.

Tourism: countries should promote environmentally sound leisure and tourism activities, making suitable use of museums, heritage sites, zoos, botanical gardens, national parks and other protected areas.

Training is one of the most important tools to develop human resources and facilitate the transition to a more sustainable world. Training for sustainable development should be focused on work, aiming to fill gaps in knowledge and skills to help people find jobs and help them to become involved in environmental and developmental work.

Building Capacity for Sustainable Development

37. The capacity of a country to develop sustainably is directly related to the capacities of its people, as well as its ecological and geographical conditions.

Nations need to assess and improve their human, scientific, technological, organisational, institutional and resource capabilities.

Capacity building is a job not only for the country involved, but for the international community as well. It is essential for individual countries to identify priorities and determine the means for building capacity and capability to implement Agenda 21, taking into account their environmental and economic needs.

Skills, knowledge and technical know-how at the individual and organisational levels are necessary for institution building, policy analysis and development management.

Technical cooperation should be directed at long-term capacity building and should be managed and coordinated by the receiving countries themselves. Technical cooperation, including technology transfer, is effective only if it is derived from and related to a country's own development strategies and priorities.

International Institutional Arrangements

38. The UN Conference on Environment and Development (UNCED), which produced Agenda 21, was the product of a UN Resolution (44/228) which mandated the conference to elaborate strategies and measures to halt and reverse the effects of environmental degradation in the context of:

    • Increased national and international efforts to promote sustainable and environmentally sound development in all countries; and

    • The recognition that the promotion of economic growth in developing countries is essential to address the problems of environmental degradation.

    Chapter 38 outlines the institutional framework and the time frame within which organs and agencies of the UN must take various actions to coordinate the implementation of the plans, programmes and projects for Agenda 21. For reasons of space, that chapter will not be condensed in this work.

International Legal Instruments and Mechanisms

39. Sustainable development will not be possible unless national and international laws and procedures are reformed to give effect to new ways of regulating the way people and nations treat their environments.

    International environmental law must be reviewed to evaluate and promote the efficacy of that legal system and to promote the integration of environment and development policies through effective international agreements or instruments, taking into account both universal principles and local needs and concerns.

Information for Decision Making

40. "In sustainable development, everyone is a user and provider of information . . . in the broadest sense. That includes data, information, appropriately packaged experiences and knowledge. The need for information arises at all levels, from that of senior decision makers at the national and international levels to the grass roots and individual levels."

In implementing Agenda 21, it is necessary to collect and organise different types of data at the local, provincial, national, regional and international levels.

Data are needed to establish the status and trends of the Earth's ecosystems, natural resources, pollution and socioeconomic variables.

Lack of appropriate data seriously hampers informed decision making. There is a general lack of capacity particularly in developing countries, but also at the international level for the collection and assessment of data and their transformation into useful information.

Among the priorities are:

- Developing useful indicators of sustainable development and using these indicators in policy making.
- Improving data collection and use by carrying out inventories of environmental, resource and development data – based on national and global priorities – for the management of sustainable development.
- Improvement of methods for data assessment and analysis, and harmonising national and international standards.
- Establishment of coordinated national and international information networks.
- Strengthening the capacity for collecting and using traditional information and for transmitting information to and from traditional sectors.
- Strengthening existing national and international mechanisms for information processing and storage to make it easier for all users to have access to the information they need, while having regard to intellectual property rights.
- Placing special emphasis on producing information in more useful forms for decision makers at all levels.
- The UN system and all national and international organisations should collaborate to identify sources of information in their organisations. The private sector should be encouraged to share its own experience.
- Countries, international organisations and non-governmental organisations should collaborate to develop and strengthen electronic information sharing (networking) capacity.

Conventions

1. Statement of Principles on Forests

Forests are essential to the maintenance of all forms of life and to economic development. Forests are sources of wood, food and medicine and are rich storehouses of many biological processes yet to be developed. They are reservoirs for water and absorb carbon which would otherwise become a greenhouse gas.

Forests are home to many species of wildlife.

They are important in fulfilling human cultural and spiritual needs.

All nations should therefore promote the planting and conservation of forests and should ensure that forests are used sustainably. International arrangements should be made to manage the forests sustainably, to meet the social, economic, cultural and spiritual needs of present and future generations. The Declaration recommends that profits from biotechnology and genetic engineering based on forest products, should be shared equitably between the developers and the national owners of forests. It recommends that special planting of forests should be made to guarantee fuel wood supplies and other raw materials, to generate income and jobs.

There should be international support for the protection of forests; and forests should be valued both for their economic and non-economic benefits.

Forestry planning should involve the entire population, especially women, youth and indigenous peoples who use them. The rights of traditional users of forests, indigenous peoples and workers should be protected.

There should be no discrimination or unilateral restrictions in trade rules governing forest products. National and international action should be taken to:

- Encourage local processing and higher prices for processed products, and,
- Regulate and control pollutants which harm forests.

2. Convention on Biological Diversity

The purpose of the Convention on Biodiversity is to protect the variety and variability of all organisms for their own sake, and because they are valuable for ecological, scientific, educational, cultural, recreational and aesthetic reasons.

Biodiversity is essential for evolution and for maintaining the life-sustaining capacity of the Earth; it is vital for food, health and other economic, social and spiritual needs of the human race.

The main aim of the Convention is to protect biodiversity by all means possible. This means that a large investment is necessary, but this investment will repay mankind with a wide and incalculable range of benefits, some far into the future.

The major principle is fair use of biological resources, including genetic material from plants, animals, microbial or any other material containing functional units of heredity, for example, DNA. The Convention recognises the need for conserving and protecting diversity as well as the necessity to preserve and protect whole ecosystems – groups of living and non-living things which operate together as a unit.

Nations should identify and monitor important components of biodiversity and develop strategies for conserving them. Biological diversity and sustainable use must become part of national and international policy planning and the development process.

Nations should educate their populations about the uses of biodiversity, establish laws to protect threatened species, promote sound environmental development and rehabilitate and restore degraded ecosystems.

Nations should also establish the means to control risks from organisms modified by biotechnology and prevent the introduction of alien species which may threaten native ecosystems. Environmental impact assessments with public participation are an important part of the policy protection necessary.

Nations should encourage people to use and to conserve their biological resources.

Countries should allow access to genetic materials within their borders for environmentally sound uses. The results of research and development of such resources must be shared fairly between the owners of the genetic material and the developers.

Developing countries are to have access to the environmentally sound technologies they need to conserve and use their biological resources and to protect biodiversity. Any technology developed to use genetic resources must be shared with those who provided the genetic resources.

Developing countries are to receive technical and financial assistance to develop their own institutions and expertise for the sustainable use of biological resources and to participate in biotechnological research. It is recommended that international regulations be developed to govern the safe handling and use of technologically modified living organisms.

Finally, developed countries that are party to the Convention have a duty to provide technical and financial aid to developing countries to help them conform to the rules of the convention. The initial funding will be handled by UN agencies involved in environment and development.

3. Convention on Climate Change

The aim of the Convention on Climate Change is to bind the world's nations to a programme for stabilising the so-called greenhouse gases in the atmosphere at levels which will not upset the global climate.

Human activities of all types have been releasing increasing quantities of gases such as carbon dioxide into the Earth's atmosphere. It is believed that these gases will cause the heating up of the Earth's atmosphere, the land and the sea, and produce a whole range of serious damage to all living creatures and the ecosystems they inhabit.

Most of the greenhouse gases are produced by the activities of the developed countries who have a special role to play in controlling this worldwide threat.

Developed countries and all countries with heavy industry, such as those in Eastern Europe, must adopt national policies to limit emissions of greenhouse gases and they must take measures to protect and improve forests, oceans and seas that act as sinks and reservoirs for greenhouse gases.

Developed countries must provide money and technological assistance to help developing countries measure the discharge of greenhouse gases, assist the most vulnerable countries with the costs they will incur to adapt to global warming, and provide environmentally sound technologies and support the development of technologies within developing countries.

All nations are required to provide information on the quantities of greenhouse gases they release and how much is absorbed by their forests, seas and other sinks.

Nations must publish regular updates on the programmes they have developed to control emissions and to adapt to climate change; and they should promote the sound management and conservation of greenhouse 'sinks' such as plants, forests and oceans.

Appendix Four

Agenda for Small Island Developing States

Small island developing states (SIDS) are recognised in Agenda 21 to be special cases, with their own severe problems. Their ecosystems are rich in biological diversity but are fragile and vulnerable both to climate change and to exploitative degradation.

Small island states are exposed to a whole catalogue of natural catastrophe, including hurricane, drought, earthquake, volcanic eruption, tsunami, saline intrusion, flooding and sea level rise due to global warming.

Economically, most are overspecialised, dependent on one or two main economic activities such as plantations, mines and tourism.

Agenda 21 recognises that small island states, most of them developing countries, must be of special concern, both for environment and development. They need international assistance to develop strategies for their peculiar developmental problems. They need to promote sustainable development within the context of their size and scarce resources.

In the spring of 1994, more than a hundred members of the United Nations were represented at the Global Conference on the sustainable development of SIDS. Also present were 14 observer territories, 53 commissions, specialised agencies and organisations and 89 non-governmental organisations from all over the world.

The conference was held in Bridgetown, the capital of Barbados, a small, developing island state in the Caribbean.

The Declaration of Barbados, adopted by the conference and endorsed by the 49th session of the UN General Assembly later that year, is the programme of action for sustainable development of SIDS.

People at the Centre

As in Agenda 21, people are at the centre of the SIDS programme of action which aims at enhancing the quality of life through policies, strategies and programmes to mobilise resources to achieve truly sustainable development.

The international community is committed to cooperate in this programme by providing adequate, predictable new and additional resources, by facilitating the transfer of environmentally safe and sound technology on concessional terms if possible and by promoting fair trade and an equitable international economic system.

Most of the resources for implementation must come from the developing states themselves, but developed countries have committed themselves to share equitably in providing financial and technical assistance to complement the efforts of the SIDS. UNCED estimates the overall costs of implementing Agenda 21 at US$600 billion annually, including US$125 billion in international grants and concessionary financing.

Principle 6 of the Rio Declaration on Environment and Development states that the special situation and needs of developing countries shall be given special priority. This principle applies particularly to the least developed and the most environmentally vulnerable states.

In Chapter 17 of Agenda 21, SIDS and islands supporting small communities are recognised as special cases for both environment and development. This is because of their environmental fragility, vulnerability, small size, limited resources, geographic dispersion and isolation. These factors are all economically disadvantageous and prevent the achievement of economies of scale in production.

The situation is complicated because the per capita income of many SIDS is higher than many developing countries – so they have limited access to concessionary resources, despite the fact that most are classified as high risk entities with limited resources.

There are 15 sections in the Barbados Declaration – the programme of action. Each section deals with a particular problem and then prescribes national, regional and international action to be taken.

In this summary we deal specifically with action to be taken nationally. Regional and international action will be summarised generally at the end of the fifteenth section.

1. Climate Change and Sea Level Rise

The survival of some SIDS is threatened by climate change. Any rise in sea levels will inundate many islands and the inhabited coastal zones of others. Loss of land will mean destruction of human settlements, contraction of exclusive economic zones and a reduction in resources. Saltwater intrusion will damage freshwater supplies. The increased frequency and violence of catastrophic weather may have profound economic effects.

National Action

SIDS should mobilise all available information on the climatic risks they face, assessing the socioeconomic implications, developing databases and information systems to monitor their situations. They must mobilise public opinion to understand the nature of the problem and what needs to be done. They need to formulate plans for more efficient use of energy and increase participation in all areas of research. They need to develop adequate response strategies.

2. Natural and Environmental Disasters

Environmental disaster is a close companion of SIDS. In fact, 13 of the 25 most disaster-prone states on Earth are SIDS. Catastrophic events due to climate change are happening more often and are more destructive. The small size and narrow resource base of the SIDS and the pervasive impact of the catastrophic events magnify and extend disaster effects on peoples, economies and environments. Costs of rehabilitation are high and insurance coverage is expensive and hard to get.

National Action

Countries must establish or strengthen disaster preparedness and management institutions and policies, develop early warning systems; establish national disaster emergency funds. They must integrate natural and environmental disaster planning into national policies and strengthen cultural and traditional systems which may improve the resilience of the communities to disastrous events.

3. Management of Wastes

SIDS are short of land space and resources. Population pressures make waste disposal and pollution prevention critical problems for SIDS. Increased urbanisation often means increased contamination of groundwater. Point-source pollution from industry and agriculture are serious factors in the degradation of the marine and coastal environment. Islands are particularly vulnerable to seaborne hazards from shipping, trade and the transport of toxic and other hazardous wastes.

National Action

National policy should strive to encourage the production and importation of environmentally friendly products. Regulations should be developed for the monitoring, reduction, prevention, control and reduction of pollution, and the safe and efficient management of toxic, hazardous and solid wastes including sewage, pesticides, herbicides, and industrial and hospital effluent.

Public awareness programmes must be developed so people can recognise the benefits of recycling, reuse and appropriate packaging. Clean technologies should be encouraged; wastes should be treated at source. Data and baseline information for waste management must be developed and used; facilities should be built for the reception of ships' waste and national laws should enforce the convention banning trade in hazardous wastes.

4. Coastal and Marine Resources

Coastal and marine resources are of vital importance to SIDS since many are entirely coastal entities. Marine and coastal zone management has been a major problem for most SIDS. Coastal habitats are increasingly degraded, marine pollution is serious and there are growing conflicts between various resource users.

National Action

Institutional, legislative and administrative arrangements must be established and strengthened. Integrated coastal zone management, including watersheds and the exclusive economic zone, must be included in national development plans.

Comprehensive monitoring programmes for coastal zones including wetlands, should determine system stability. Traditional knowledge and management practices should be integrated if ecologically sound; local communities must be involved in the management of the coastal zones.

Nations must develop capacities for the sustainable harvesting and processing of marine resources and provide training and awareness programmes for governments and local communities involved in coastal zone management.

5. Freshwater Resources

Freshwater resources are vital to all countries, not least because they set limits to development. In developing countries, poor freshwater quality causes disease, while increased urbanisation endangers freshwater supplies. Island freshwaters are threatened by deforestation and soil erosion, siltation, seepage, pollution from solid and liquid wastes, including human wastes, animal wastes, industrial wastes, fertilisers, pesticides and herbicides.

Development prospects are threatened because watersheds and groundwater resources are not well protected. Enormous capital expenditures may be needed to overcome these deficiencies.

National Action

States should develop integrated water plans, using appropriate incentives and regulations to develop, maintain and protect watershed areas, irrigation systems and distribution networks. They should promote water conservation and prevent pollution by programmes which involve communities in management and conservation of watersheds and forests, in reforestation and in devising investment strategies.

States need to implement comprehensive water planning in order to be able to forecast supply and demand. They should strengthen procedures to monitor the effects on water supply of climate change and sea level rise and make plans to anticipate these crises. They should encourage the development of appropriate technology to devise low-cost and cost-effective sewage disposal, desalination and rainwater collection to provide adequate quantities of clean drinking water. States should develop capacities to rationalise the allocation of scarce water supplies among competing sectoral interests.

6. Land Resources

Land is a scarce resource in most SIDS. Land tenure systems, soil types, geography and climate also tend to limit the area available for urban settlement, agriculture, mining, commercial forestry, tourism and other infrastructure. There is intense competition for land. More efficient and just ways of using these resources must be developed.

The major issue is degradation because of overuse of the land. Catastrophic events, earthquakes, volcanoes and hurricanes may also contribute to degradation. Large-scale developments often mean the marginalisation of poor farmers and the acceleration of the decline in land stability and fertility. Groundwater quality, river quality and marine and coastal zone amenity are also damaged because of siltation and pollution of reefs, lakes and lagoons.

National Action

Good public and official information is crucial in land management. Carrying capacities of land must be determined. Rational decision making is possible only if the economic and environmental value of land resources is known. Local communities, particularly the youth and women, must be brought into the decision making processes.

Comprehensive land use and zoning plans and land pricing policies must be developed and implemented, to sustain productive land use and prevent degradation. Appropriate forms of tenure are to be encouraged, backed by appropriate new laws and regulations which must be enforced. Watershed and coastal zone protection must be community responsibilities, involving everyone, including large landowners, in afforestation, reforestation and the promotion of natural regeneration.

Shelter quality must be improved and there must be increased reliance on national physical planning in rural and urban environments, focusing on training to strengthen physical planning and using environmental impact assessments and other decision making tools to involve the communities in the process.

7. Energy Resources

Most SIDS are heavily dependent on imported energy, mainly petroleum, for transport and power generation. In some states, imported energy accounts for more than 12 percent of imports. Many will continue to be dependent on indigenous biomass fuels for cooking and crop drying.

All SIDS have substantial, but still underdeveloped, solar energy prospects. Many have substantial wind power potential; hydroelectric power is available only in a few. In most SIDS, fuel is used inefficiently and wastefully. At the same time, there are hindrances to the large-scale commercial use of renewable energy resources, mostly because of technological and investment shortages, as well as shortages of management and other skills.

Renewable energy resources can be developed for commercial use only if appropriate technologies and training are available.

National Action

Public education and awareness is vital to sustainable development in the energy sector. States must promote the efficient use of all energy and develop environmentally sound and safe sources of renewable energy, using, if necessary, economic instruments as incentives for these developments.

States need to develop or strengthen research capabilities into the development and promotion of new and renewable sources of energy including solar, wind, geo-thermal, hydroelectric, wave (seawater), and biomass energy and ocean thermal energy conversion. They must strengthen research capabilities and develop technologies to encourage the efficient use of non-renewable sources of energy.

8. Tourism Resources

Tourism continues to be a substantial contributor to the economies of many SIDS. However, if not properly planned and managed, tourism could significantly degrade the environment on which it is so dependent.

The overdevelopment of tourism can be environmentally and socially destructive and disruptive to entire communities and, sometimes, to whole islands.

National Action

States should ensure that tourism development and environmental management are mutually supportive. Tourism development should be the result of integrated planning, particularly of land use and coastal zone management. All tourism projects should require environmental impact assessments and all should be continuously monitored for environmental effects.

Tourism should be regulated by guidelines and standards for water and energy consumption, waste generation and disposal, land degradation. Regulations and guidelines should also be established to deal with the proper management and protection of ecotourism attractions, and the carrying capacity of tourism areas.

Specific facilities should be identified and developed for niche markets, particularly in ecotourism, nature tourism and cultural tourism. These projects should involve local populations in the identification and management of natural areas set aside for ecotourism.

States should protect their cultural integrity.

9. Biodiversity Resources

Small islands are renowned for species diversity and endemism. At the same time, their small size, isolation and vulnerability to disasters, added to population pressure, make their ecosystems among the most threatened on Earth. Deforestation, coral reef deterioration, habitat destruction and degradation, and the introduction of exotic species are all serious threats to their biodiversity.

Although there is usually enough information available for decision making, many sites requiring *in situ* conservation have been neglected and left unprotected. In many states, traditional land and resource ownership requires that conservation must be a community effort. This means that conservation strategies must take into account traditional systems of land tenure and use.

National Action

Integrated strategies must be developed for species conservation and sustainable use of biodiversity. There must be protection from exotic species. Areas of high biological interest need to be identified and protected.

Education at all ages should increase awareness of the values of conservation of biodiversity and the designation of protected areas. Resource owners must be made aware of the fundamental importance of maintaining a diverse biological base.

Buffer stocks and gene banks should be established and maintained for reintroduction of biogenetic resources into their natural habitat, especially important after disastrous events. There should be increased emphasis on study and research into biological resources, their management and intrinsic cultural and socioeconomic values, including biotechnological potential.

Detailed inventories should be made of existing flora, fauna and ecosystems to provide basic data for biodiversity conservation. Ownership of intellectual property rights must be protected and the technology, knowledge, customary and traditional practices of local and indigenous people must be effectively protected so that they benefit directly and equitably on mutually agreed terms from any direct or derivative use of their technology, knowledge or practice.

Community and other non-governmental organisations and local women, indigenous peoples and other major groups, including fishermen and farmers, must be involved in the conservation and sustainable use of biological resources.

10. National Institutions and Administration

The interrelatedness of all activities is most strongly marked in small island states because of their small size. Environmental considerations must be an integral part of national decision making and is considered to be the single most important step in the achievement of sustainable development.

The principle of sustainability must be seen to guide all future development.

National Action

The capacity to implement Agenda 21 depends on the strengthening of institutions and administrative capacity. Cross-sectoral and interministerial committees and task forces should be used to integrate planning across all sectors.

Sustainable development depends on the development of strategies and schedules for financing and strengthening environmental agencies.

Communities and non-governmental organisations must be involved in decision making, in public education, and the implementation of programmes. Public education must be improved to make all relevant bodies and authorities aware of their responsibilities and of the environmental legal framework.

National policy should facilitate public discussion on the value of environmental legislation and standards to the community as a whole. There should be the widest and most open discussions on decisions regarding appropriate penalties for contravention of environmental laws and regulations. Regulations at national and lower levels must reflect the needs and incorporate the principles of sustainability and ensure public participation in environmental impact assessment for both public and private sector development.

States should provide adequate resources and authority for the proper functioning and enforcement of environmental regulations. Domestic legislation should bring national policy into line with international conventions and agreements to which the state has acceded.

National information nodes on sustainable development must be established to encourage the development of international networks of experience, information and data on sustainable development for SIDS.

11. Regional Institutions and Technical Cooperation

Regional organisations, both of the UN and from outside, have key roles to play in the achievement of sustainable development. They will facilitate efficient and effective assistance to SIDS and will be used to implement and coordinate regional programmes.

Currently, multilateral and bilateral donors undertake their own regional programming exercises.

National Action

Regional bodies should improve coordination of regionally formulated programmes and strategies, and should cooperate and assist national bodies. Regional bodies should be used to develop a SIDS Technical Assistance Programme (TAP) to promote inter- and intra-regional cooperation in sustainable development.

Regional development centres should be established to facilitate the coordination of research, training, the development of indigenous technology, the transfer of technology and the provision of technical and legal advice.

Regional institutions should be responsible for preparing environmental law training manuals for lawyers and non-lawyers in the environmental field. They should conduct regional and national workshops on environmental law, impact assessment, heritage, pollution, and environmental mediation. The regional institutions have the responsibility for keeping SIDS informed about the content, notification processes, financial and legal implications of relevant international environmental instruments and to help them accede to and implement them.

12. Transport and Communication

Islands depend on reliable transport and communication lifelines for contact with the rest of the world. Most islands have no control over their international communications or transport. The adequacy and pricing of these services is also outside of their reach. Domestic markets are too small and often too remote to supply the economies of scale necessary for the most efficient operation.

In addition, land transport has proven to be one of the most dangerous degraders of the urban environment. Innovative ways need to be designed to make transport more efficient and to serve populations better, more efficiently and more economically.

National Action

SIDS should continue efforts to strengthen transport services and facilities, while ensuring environmental protection and safety. Energy-efficient and innovative low-cost solutions can be developed. Domestic communications facilities, including radio broadcasting and telephony must be upgraded to serve the greatest number, including those in the most remote areas.

Small states should move towards increased cooperation and consolidation of their domestic airlines and plan communications development on a regional basis. They should also promote the application of appropriate communications technologies to promote sustainable development in areas such as education, health, ecotourism and other areas critical to sustainable development, including the promotion of greater public awareness.

13. Science and Technology

The achievement of sustainable development depends on the wise use of science and technology. In many small states, science and technological capacity remain undeveloped. While some island people survive on traditional knowledge, SIDS are increasingly being driven to abandon traditional ways, even if they are useful, to adopt modern technology. What is necessary is a better integration of modern technology and traditional knowledge and practice.

National Action

Sustainable development should link science and technology to national environmental strategies, and be responsive to local and sectoral plans, emphasising self-sufficiency and the minimisation of import dependency.

Greater emphasis should be put on research and development, training for science and technology, and economic development generally, and for environmental and technology assessment in particular. Natural resource accounting should be enhanced by the refining of analytical tools. Information and communications technology should be adapted to overcome problems of size and isolation.

There should be more research into indigenous, traditional practices in areas such as agriculture, agricultural processing, waste recycling, ethnobiology and biotechnology, construction and renewable energy. Indigenous, environmentally friendly technologies should be encouraged by establishing standards, incentives and regulations for their use. It is important to strengthen the role of women in science and technology disciplines.

14. Human Resource Development

Since human beings are at the centre of sustainable development, strategies should emphasise projects which enhance the quality of life. Projects should be undertaken with a view to maximising the contribution that individuals, groups and communities can make towards sustainable development and to maximise the well-being of people.

National educational and training mechanisms must be strengthened as a matter of high priority in order to widen the flow of information on sustainable development issues. People should be made aware of the environment in which they live and be given the opportunity to manage it.

National Action

The idea of sustainable development must be infused into educational curricula at all levels. It should be national policy to encourage public participation in environmental responsibility and management and to encourage the understanding that environmental, social and economic issues are all interconnected.

There should be greater access to mathematical, scientific and technical training.

Population issues must be incorporated into the mainstream of decision making and planning.

Planning should promote the dignity and fundamental rights of the human person and the family.

It should be national policy to improve all human settlements, in consultation with the communities themselves, giving priority to the provision of basic services. These services should include access to potable water, environmentally sound and safe sewage treatment and disposal, shelter, education, health, family planning and the elimination of poverty.

Development projects should be people-centred and have explicit health and environmental objectives. They should ensure the provision of adequate resources for public health and preventative medical activities. They should take into account urban development options, including decentralisation.

Nations should raise standards of human settlements by designing projects aimed at eliminating poverty. National policy should encourage the development of distance teaching and training to meet the large demand for knowledge and training in the area of the environment.

Major groups, including non-governmental organisations and women, should be encouraged to participate fully in the creation and implementation of sustainable development initiatives.

Traditional and indigenous knowledge and skills should be treasured, especially in the area of resource management and health. Environmental awareness should be promoted through community groups.

15. Implementation, Monitoring and Review

Effective implementation of the programme of action depends on the effective monitoring and review of the programme.

The SIDS Programme of Action provides the international community with a chance to demonstrate its stated commitment to implementing the principles of Agenda 21.

Since SIDS are particularly environmentally vulnerable, the United Nations system and the international community, in line with principle 6 of the Rio Declaration and on the basis of Chapter 17 of Agenda 21, shall give special priority to the situation and needs of SIDS.

That will require, in particular, providing adequate resources for the implementation of the programme and for action at national, regional and international level.

The implementation of the Programme of Action will proceed in parallel with international processes which are part of the agenda for sustainable development. These processes include commissions, conventions and protocols dealing with specific issues, including climate change, biodiversity, desertification, the movement of hazardous wastes, trade in endangered species and a range of international agreements to safeguard the environment and promote sustainable development.

National Action

Although SIDS have begun to implement Agenda 21, there is need for more work and faster progress, particularly to ensure that environmental considerations are given full significance in national plans, at both micro and macro levels.

Progress on the programme of action will ultimately depend on the resources mobilised by and made available to SIDS from internal and external sources, particularly in the area of building institutional capacity.

Critical to the effective implementation of the objectives, policies and mechanisms agreed to by the governments in all programme areas of Agenda 21 will be the commitment and genuine involvement of all social groups. New participatory approaches to policy making and implementation of sustainable development programmes will be necessary at all levels.

In that regard, there are essential roles for women, youth, senior citizens, indigenous people and local communities, as well as the private sector, labour and non-governmental organisations. As stated in Agenda 21, one of the fundamental prerequisites for the achievement of sustainable development is broad public participation in decision making.

Finance Although financing for the programme of action will come from the public and private sectors of the SIDS, innovative financing options need to be explored to deal with the special circumstances of the SIDS.

It should be national policy to maximise the impact of available resources and to explore the use of economic instruments; to promote private sector investment, using innovative financial mechanisms including small scale grants and loans for micro enterprise at the community level.

Official Development Aid (ODA) will be a main source of funding for the least developed SIDS.

Trade SIDS should seek to diversify their production structures for goods and services that exploit existing or potential comparative advantages.

Technology National policy should specially encourage the collection of information and development of indigenous technology and the development of management capacities for assessing, acquiring, disseminating and developing appropriate technologies, while adequately protecting intellectual property rights.

Legislation New legislation should be developed as needed, incorporating customary and traditional concepts where appropriate, backed by training and adequate resources for enforcement.

Institutional development   National policies should ensure the integration of environmental, population and development strategies in national and sectoral planning.

Information and participation   A prime objective of Agenda 21 is to increase the awareness and involvement of non-governmental organisations and women, youth, local communities and other major groups in national planning, in developing environmentally sound and sustainable technologies and in implementing sustainable development activities.

Human Resource Development   Public awareness and human resource development will promote national awareness at all levels, including education and training, skill development, particularly of technicians, scientists and decision makers, to enable them to better plan and implement sustainable development.

Appendix Five

Important Dates for Sustainable Development

1896	Arrhenius predicts global warming
1957	International Geophysical Year finds evidence of global warming
1962	Rachel Carson's *Silent Spring* published
1972	UN Conference on the Human Environment, Stockholm Establishment of UN Environmental Programme – UNEP
1975	Second World Conference on Women, Mexico City
1976	First UN Conference on Human Settlements – Habitat I, Nairobi, Kenya
1983	UN Commission on Environment & Development (Brundtland Commission) set up
1987	*Our Common Future* (Report of Brundtland Commission) published Montreal Protocol to Protect the Ozone Layer.
1992	UNCED: The UN Conference on Environment and Development, Rio de Janeiro, Brazil – AGENDA 21 signed
1993	World Conference on Human Rights, Vienna, Austria
1994	UN Global Conference on the Sustainable Development of Small Island Developing States, SIDS, Bridgetown, Barbados World Conference on Natural Disaster Reduction, Yokohama, Japan
1994	International Conference on Population and Development, Cairo, Egypt

1995   World Summit for Social Development, Copenhagen, Denmark
        Fourth World Conference on Women, Beijing, China

1996   UN Commission on Sustainable Development – Initial review of
        Agenda 21
        Second UN Conference on Human Settlements – Habitat II,
        Istanbul, Turkey

1997   UN Conference on International Migration and Development
        UN Commission on Sustainable Development – Overall review
        of Agenda 21 by Special Session of the General Assembly

1999   Second Global Conference on the Sustainable Development of
        Small Island Developing States, SIDS
        UN Full Review of SIDS Programme of Action

Index